Cambridge Primary Revise for Primary Checkpoint

World English

Jennifer Peek

Teacher's Handbook

HODDER
EDUCATION
AN HACHETTE UK COMPANY

Acknowledgements

The Publishers would like to thank the following for permission to reproduce copyright material. Every effort has been made to trace or contact all copyright holders, but if any have been inadvertently overlooked, the Publishers will be pleased to make the necessary arrangements at the first opportunity.

Photo acknowledgements

p. 26 *cl* © Solaru S/Adobe Stock Photo; **p. 26** *cc* © Antonio/Adobe Stock Photo; **p. 26** *cr* © Gemini/Adobe Stock Photo; **p. 26** *cc* © MP P/Adobe Stock Photo; **p. 26** *cl* © Rehtse C/Adobe Stock Photo; **p. 26** *cr* © Monticellllo/Adobe Stock Photo.

t = top, *b* = bottom, *l* = left, *r* = right, *c* = centre

This text has not been through the Cambridge International endorsement process.

Although every effort has been made to ensure that website addresses are correct at time of going to press, Hodder Education cannot be held responsible for the content of any website mentioned in this book. It is sometimes possible to find a relocated web page by typing in the address of the home page for a website in the URL window of your browser.

Hachette UK's policy is to use papers that are natural, renewable and recyclable products and made from wood grown in well-managed forests and other controlled sources. The logging and manufacturing processes are expected to conform to the environmental regulations of the country of origin.

Orders: please contact Hachette UK Distribution, Hely Hutchinson Centre, Milton Road, Didcot, Oxfordshire, OX11 7HH. Telephone: +44 (0)1235 827827. Email education@hachette.co.uk. Lines are open from 9 a.m. to 5 p.m., Monday to Saturday, with a 24-hour message answering service. You can also order through our website: www.hoddereducation.com

© Jennifer Peek 2022

First published in 2022 by

Hodder Education
An Hachette UK Company
Carmelite House
50 Victoria Embankment
London EC4Y 0DZ

www.hoddereducation.com

Impression number   10 9 8 7 6 5 4 3 2 1

Year   2026 2025 2024 2023 2022

Cover by Lisa Hunt from the Bright Agency

Illustrations by Vian Oelofsen

Typeset in FS Albert 11/13 by IO Publishing CC

Printed in the UK

A catalogue record for this title is available from the British Library.

ISBN: 9781398369887

Contents

Introduction

What is this book about?

The aim of the Teacher's Handbook is to help teachers work alongside learners who are revising for the Cambridge Primary Checkpoint tests. It accompanies the *Cambridge Primary Revise for Primary Checkpoint World English Study Guide*, which is a write-in revision aid for learners. This Teacher's Handbook clearly explains the revision exercises and activities in the Study Guide and suggests ways that these could be extended or differentiated to provide further support. Notes about misconceptions and things to look out for, as well as answers to the exercises in the Study Guide, are also provided in this Teacher's Handbook.

The Teacher's Handbook is divided into four main revision units, mirroring the units in the Study Guide: *Listening and speaking*, *Writing*, *Reading* and *Use of English*. Each unit is split into sections covering smaller subtopic areas, which are based on the learning objectives of the Cambridge Primary World English Framework. Each section ends with *Let's revise*, which provides linked revision ideas and a *Top tip*, which gives extra teaching ideas or notes for you to implement in the classroom. The *Let's Go* feature at the end of each section in the Teacher's Handbook includes a photocopiable worksheet that can be used for further revision or reinforcement of the language concepts.

Although the Study Guide is a write-in book, it will be useful to have other materials available for learners to use, including rough paper or notebooks, colouring pencils or pens, highlighters, whiteboard markers and sometimes internet access.

Why is revision important?

Revision gives learners the opportunity to reinforce and reflect on what they have learned. It also helps learners to identify what they are secure with and what they do not know, and then to work on what they have forgotten or remain uncertain about. Revising means that language learning is extended and deepened. It is also a very real and tangible way for learners to see just how far they have progressed along their language-learning journey. Revision is not just about revisiting concepts to prepare for a test. Above all, it is a self-reflection exercise. It is an opportunity to discover strengths and identify areas of uncertainty. As such, it should not be approached as a limited, one-method-suits-all process, because all learners have different strengths and weaknesses.

Revision should be a collaborative process between learners and teachers. Emphasise that it is okay to make mistakes, especially when revising. When working through the Study Guide with learners, frequently ask them to provide extra examples of language points. Encourage learners to always be honest with themselves and with you as their teacher. Make sure that they understand it is okay to acknowledge when they are not sure about something and to seek help or clarification. Sometimes, even after additional support, a learner may still not fully understand something, such as a grammar point, in which case you might need to go back a few steps in their learning to help them unpack what it is that they don't understand, or to undo any misconceptions that may have formed. In this way, you can produce bespoke explanations or examples for individuals to help move their learning forward.

Unit 1 · Listening and speaking

Section 1: Personal information

 Study Guide pages 5–7

 Audio 1.1, 1.2

 Key skills: Recognising question words; identifying main points; understanding details; making notes; giving a short talk.

Key words and question words

Study Guide (pages 5–7)

Learn

- This information explains the function of key words and question words in English. Before reading the definitions with the class, ask learners to share their ideas about what key words and question words do.
- Ask volunteers to tell you any question words that they know and write them on the board. It may be useful to tell learners to think about the 'w' questions: *who, what, when, where* and *why*.

Try this

- This is a listening exercise.
- Play *Audio 1.1* from the online resources at www.hoddereducation.co.uk/cambridgeextras. Learners listen to a conversation between two people and make notes to answer questions about it in a table. The notes relate to the 'w' questions. Learners need to identify who is speaking, what they want to do, why they want to do it, when and where they will do it, and any other information.
- It is always useful to play the audio recordings for listening exercises at least twice. Tell learners to read the questions before playing the audio recording. Explain to learners that they should not try to write anything at all on the first listen. Then, on the second listen, they can make notes.
- Play the audio recording again for a third time, or more for extra support, and allow time at the end for learners to look at their notes, and add or change anything before going through the answers as a class.
- When reviewing the answers, you can play and pause the audio recording in relevant places to highlight the information for consolidation. You may wish to complete the next *Practise* activity before going through both sets of the answers, as both activities relate to this audio recording.

Answers

Question	Answer
Who is speaking?	A boy (Sesh Goudry)
What do they want to do?	Enter a race / a swimming race / the Meadow Bridge race
When will this happen?	10 November (tomorrow) 8 a.m.
Where will this happen?	Meadow Bridge
Why do they want to do it?	To show people they are as fast as any adult
What other information did you hear?	Any answers from: Sesh is 12 years old. The entry fee is $10; Sesh saved the money himself. Sesh gets up at 6 a.m. every morning. If he is late, he won't be allowed to swim. Sesh thinks he will be in the top three. The adult wishes Sesh good luck.

Practise

- This is a reading and responding task, based on the information in the listening exercise from the previous *Try this* activity.
- Learners read questions and choose the correct response from the multiple-choice options provided. Advise learners to read the questions and options slowly and carefully because it is very easy to make mistakes when rushing.
- Some of the multiple-choice options here differ only in very small ways, so learners need to pay close attention when they are reading.
- Tell learners to read the advice in the *Hint* box. Remind them to focus on the main points as they listen to the audio recording. They must not get distracted by the details.

Answers

1 A	2 B	3 C	4 C	5 B

Let's talk

- Learners should complete this speaking activity with a partner. In pairs, learners practise asking and answering questions politely in conversation with each other. A conversation structure is provided and learners decide on the missing details to complete the conversation.
- You could ask pairs to share their conversations in small groups or as a class. While learners are working on their conversations, circulate around the class and listen for learners who are collaborating well together and demonstrating confidence with spoken English. Encourage learners who are less confident to keep trying and reassure them that they are doing well.

Practise

- This is a writing exercise. Learners complete the missing details to give personal information about themselves including their name, age and where they live, as well as what they like to do and something interesting about themselves.
- Tell learners to use neat, legible handwriting in their written work. Remind them that sentences start with an upper-case letter, end in a full stop and have appropriate punctuation such as commas and apostrophes in between. It may be helpful to review basic punctuation rules with less-confident learners.

Practise

- This is a listening comprehension. It is always useful to play the audio recording for listening exercises at least twice. Tell learners to read the questions before you play the audio recording.
- Play *Audio 1.2* from the online resources at www.hoddereducation.co.uk/cambridgeextras. Learners will listen to a conversation between two people and respond by choosing the correct option to answer five questions. The *Hint* box encourages learners to listen for the names of people and places in the conversation, to look at all of the options and to choose the answer that they hear, instead of guessing.
- Ask the class what else they can tell you about the names of people and places. What types of words are they? How do we recognise these words when we read them? These words are proper nouns, which are nouns that begin with an upper-case letter.

Answers

1 B	2 A	3 C	4 B	5 A

Try this

- This note-making activity could be carried out in class or it would be suitable as a homework activity with a follow-up task in class.
- For the first part of the activity, learners carry out some research about a sports or music personality that they like. They make notes in the table about where the person comes from, the language they speak, what they do, why they like them and anything else they feel is interesting. Learners can use books, the internet, magazines and newspapers, or even podcasts to do research about their chosen person.
- For the second part of the activity, learners use the notes they have made to give a short talk about their chosen person. Learners could do this with a partner, in small groups or as a class.

Let's Revise

- Revise proper nouns for countries and cities by using a world map, atlas or globe.
- Hold a spelling bee. Ask learners to spell words for countries and cities aloud letter by letter. How many will they remember?
- Challenge learners to plan and present a profile about a classmate.

Top tips

Encourage learners to use English as often as they can in class. Greetings, goodbyes, please, thank you and basic questions should always be asked and responded to in English.

Let's go! Worksheet 1: Personal information

- Worksheet 1 provides further reinforcement and practice in giving personal information. This is a continuation of the *Practise* activity on page 6 of the Study Guide, but here learners imagine it is the future and they are writing about their past. They can use the template on page 6 as a guide.
- Tell learners that they can adapt and change sentences to make them work in the past and they can add extra sentences of their own if they wish to as well.
- When learners have written their paragraph, put them into pairs or small groups to practise giving the information by reading their work aloud. Learners could then give each other constructive feedback and advice, such as encouraging each other to speak a little louder or more slowly when reading aloud.

Worksheet 1

Personal information

Extra challenge: *Practise*, page 6

1   Imagine that you are fifty years old and you are talking about what you were like as a child. Use the template on page 6 to help you to write the information, but this time use the past tense. Include details about your age, where you lived, and your likes and dislikes. Write your paragraph in the space below.

I'm going to add extra information about where I went to school!

_____

_____

_____

_____

_____

_____

_____

_____

_____

_____

_____

_____

_____

_____

_____

_____

2   Now practise giving this information by reading to a partner or a small group.

Section 2: Instructions and directions

 Study Guide pages 8–11

 Audio 1.3, 1.4, 1.5

 Key skills: Sequencing directions; understanding and giving instructions; identifying main points and ideas.

Giving and following instructions and directions

Study Guide (pages 8–11)

Learn

- Ask the class for volunteers to give you an example of an instruction they know. Encourage learners to think about instructions they hear every day at home, in school, or in other places.
- Use the information in the *Learn* box to remind learners that instructions are all about *how* to do something. Instructions are always in the present simple tense and connectives tell us the order in which to follow the instructions.

Try this

- This fun listening activity revises language for directions. Play *Audio 1.3* from the online resources at www.hoddereducation.co.uk/cambridgeextras. Learners will listen to Pia giving instructions to get to her house while they follow her route on the map provided.
- There are further activities that can be done using the map to extend this activity and include a wider variety of language concepts and skills. For example, learners could label or write a list all of the places and features on the map, such as the houses, Pia's house, the bank, the hospital, the school, the train station, trees, hills, traffic lights and the lake.
- Worksheet 2 at the end of this section provides more revision of directions and uses the map from this activity.

Practise

- These multiple-choice questions are based on the map from the *Try this* activity on page 8. Learners could answer these questions individually, or less-confident learners could work with a partner for support.

Answers

1 C   2 B   3 B   4 C

Practise

- These questions revise the use of connectives to join parts of sentences together. First, learners choose the correct word to complete sentences and then they choose words to fill gaps in a paragraph.
- Remind learners that instructions about how to get somewhere are called directions, as explained in the *Hint* box.

Answers

1 a until   b when   c next to

2 **First**, walk straight down the road **until** you come to a set of traffic lights. **Next**, turn right at the traffic lights. **Then** carry straight on through the next set of traffic lights.

Let's talk

- This speaking task gives learners the opportunity to construct their own instructions by writing directions to get from school to home. First, learners write a plan in the space provided. A plan can be rough notes, or a flow chart, or even pictures to help them to remember the directions. Learners can then write their sentences in their notebooks or on paper and, when they are finished, they can share their directions with a partner, in a small group or with the class.

- Tell learners to read the bullet point instructions before they begin. They should remember to use the present simple tense, to keep directions short and to use connectives to show the order in which the directions are to be followed.
- Listen to learners' directions, give constructive feedback to support learners who make mistakes and provide praise to all learners for taking part and trying their best.

Learn

- This part of Unit 1 revises listening for the main points. Explain to learners that listening for main points means not trying to understand every word, or fine details. Instead, it means to listen for the general ideas or for key information.
- Ask two volunteers to read the speech bubbles to emphasise this.

Practise

- This is a listening exercise about understanding the main points. Play *Audio 1.4* from the online resources at www.hoddereducation.co.uk/cambridgeextras. Learners will listen to information about a competition and choose the correct picture option to answer each multiple-choice question. They are listening for the key facts: what they need to make, where the race starts and at what time the race starts.
- Encourage learners to listen carefully for those key points and not to try to understand every word they hear. Additionally, before or after carrying out the activity, you could make sure that learners know the vocabulary for all of the picture options for some consolidation of language.

Answers

1 C   2 A   3 C

Try this

- This is a fun task with multiple activities based on an audio recording.
- First, play *Audio 1.5* from the online resources at www.hoddereducation.co.uk/cambridgeextras. Ask learners to listen carefully for the ingredients. Leaners could write down the ingredients that they hear (there are seven in total). Encourage learners to look at the pictures – can they name the ingredients shown on page 11?
- Next, learners find the seven ingredients hidden in the word search grid. The *Hint* box offers some advice about the word search puzzle for learners who may be unfamiliar with this type of puzzle.
- Finally, in pairs, learners have a conversation about how to make pancakes. They can talk about the information they heard in the audio or their own experiences of making pancakes. It may be helpful to play the audio recording again before learners carry out the speaking activity.

Answers

s	r	t	e	z	x	k	l	w	y	n	e
u	a	c	m	l	p	f	x	e	n	w	m
g	k	l	v	k	f	l	o	u	r	s	i
a	o	u	t	w	x	z	r	i	m	n	l
r	k	c	b	p	e	o	g	k	l	q	k
n	y	w	l	k	g	x	z	e	t	r	s
b	a	k	i	n	g	p	o	w	d	e	r
i	s	r	z	o	b	w	k	v	l	k	g

Let's revise

- Ask learners to say the times of day when they do different things such as having breakfast or going to sleep.
- Set up an obstacle course in the school grounds. Put learners into pairs and tell them to take turns to guide their partner around the obstacle course by giving directions. Make this more challenging by having one partner close their eyes tight shut or using simple blindfolds – but be sure to risk assess the space if you do so!

Top tips

- To provide differentiated support for more confident learners, you could talk about how the verbs used in instructions are usually in the imperative form.
- You could also talk about how we can use adverbs to add more detail to instructions, for example, 'walk slowly'.

Let's go! Worksheet 2: Give and follow directions

- Worksheet 2 provides further reinforcement and practice in giving directions. Learners use the map from page 8 in the Study Guide and work with a partner to give directions from one place to another.
- There is an extra team challenge for learners to give directions to every house on the map. Tell learners to read the advice in the speech bubble. They need to use ordinal numbers so that we know exactly where the directions end since the houses are in rows of more than one. Using ordinal numbers is also their only way of identifying houses as the map is monotone and not in colour. For example, they could end with '… and it's the second house after the traffic lights.' You may need to support less-confident learners by reminding them of ordinal numbers before they complete this activity. This activity could be extended for more confident learners by challenging them to give directions between any two or more numbered houses or labelled places on the map.

Worksheet 2

Give and follow directions

1   Work with a partner. Look at the map. Then take turns to give the following directions:
   - from the train station to the hospital
   - from the school to the lake
   - from the bank to the school
   - from the lake to the train station.

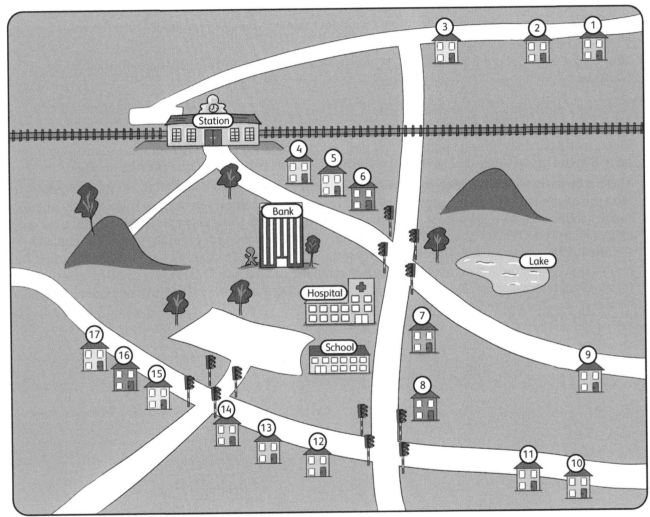

2   Work in groups to complete this team challenge. Start at House 1 in the top right corner of the map. Give directions to get from that house to every other house on the map.

We need to use ordinal numbers to say which item when we talk about more than one!

Section 3: People, places and happenings

 Study Guide pages 12–15

 Audio 1.6, 1.7

 Key skills: Identifying main points; role playing; using the present tense; listening for details; identifying and pronouncing adjectives and adverbs.

Listening for the main points

> **Study Guide** (page 12)

Learn

- This section develops and consolidates listening for the main points in spoken English. The speech bubble in the *Learn* box gives some further advice about what to listen for in order to quickly identify the main points.
- Emphasise that listening for the main points means listening for what is important and not worrying about extra information that isn't needed.

Practise

- This is a listening comprehension. It is always useful to play the audio recording for listening exercises at least twice. Tell learners to read the questions before playing the audio.
- Play *Audio 1.6* from the online resources at www.hoddereducation.co.uk/cambridgeextras. Learners will listen to Ali and Ben talking about their school trip. Learners need to pick out some key information about where the boys are, the places they visit and what they think about the places.
- Point out the speech bubble before carrying out the activity. This reminds learners to listen out for connective words that we use to sequence events (to put them in order).

Answers

1 C   2 B   3 A   4 A

Let's talk

- Ask learners to do this speaking activity in pairs or small groups. Learners work together to construct and read a role play about going on a school trip. They should talk about the places they will visit and should give their opinions about them.
- Pia's speech bubble gives an example that learners can use as a starting point. Remind learners that when we are describing places and giving our opinions, we should use present tense verb forms, as explained in the *Hint* box.

Listening for detail

> **Study Guide** (pages 13–14)

Learn

- Explain to learners that listening for details is about listening for specific information. Details are extra pieces of information that tell us more about the main points.
- Learners will go on to develop their skills in listening for details in the *Try this* activity when they will listen to the traditional story, *The clever fox and the greedy crow*.

Hint

- This information tells learners more about traditional stories. You may wish to spend some time talking about or reading some traditional stories from learners' countries or from cultures around the world.
- Traditional stories often include characters such as animals that can talk and they usually have a message or moral – something that can be learned by listening to the story.

Try this

- Play *Audio 1.7* from the online resources at www.hoddereducation.co.uk/cambridgeextras. Ask learners to listen to this audio recording of the traditional story, *The clever fox and the greedy crow.*
- On the first listen, tell learners to sit still and quietly listen without doing anything else.
- On the second listen, explain to learners that they can make notes to respond to questions 1–4.
- Point out to learners that they need to give more than one-word answers or short phrases. The point of the activity is to listen for details, so they should give plenty of information to answer each question. It may be helpful to play the audio recording at least three times to give learners ample time to write their responses.
- Learners can then share, compare and discuss their responses. You could talk through the answers as a class, playing the audio recording again, to help them hear anything they may have missed.

Answers

1 The fox and the crow.
2 The fox is clever, and the crow is greedy.
3 The fox takes the crow's cheese.
4 Don't be fooled by flattery.

Practise

- This is a reading and responding task, based on the information in the listening exercise from the previous *Try this* activity. Learners read questions and choose the correct response from the multiple-choice options provided.
- Advise learners to read the questions and options slowly and carefully because it is very easy to make mistakes when rushing. Some of the multiple-choice options here differ only in very small ways, so learners need to pay close attention when they are reading.

Answers

1 C	2 B	3 A	4 C	5 A

Practise

- This final activity, based on the story of *The clever fox and the greedy crow,* has several questions for learners to complete.
- First, learners read sentences about the story and choose the word or phrase that will make the sentence correct. To support less-confident learners, you may wish to play the audio again before learners answer question 1.
- Next, they work with a partner to practise the pronunciation of some adjectives and adverbs they have heard in the story. Learners need to put the stress (emphasis) on the correct part of each word, which is shown in bold.
- Finally, learners complete a writing activity where they write sentences of their own. They write a sentence for each of the adverbs and adjectives from question 2. Remind learners to use neat handwriting and to punctuate their sentences appropriately.

Answers

1 a clever
    b greedy
    c strong and beautiful
    d as sweet as a songbird's
    e stupid
2 Learners' own responses.
3 Learners' own responses.

Telling a story

Study Guide (page 15)

Do you remember?

- There are many different forms of storytelling. What can learners tell you about the features of a story?
- Ask volunteers to read the speech bubbles, which describe some features that all stories have.

Let's talk

- On page 15, learners see a set of six pictures that illustrate the traditional story of *The Tortoise and the Hare*.
- They will use the pictures to tell the story to their partner. In order to help them to do this, learners need to first make a plan by noting down some of the key details they need to include. Learners make notes about the names of characters, the setting, the basic plot, the ending and what message or moral the story has.
- When they have noted all this information, they can use it to retell the tale to a partner in their own words. The *Hint* box provides some adverbs that will be useful. They can use these adverbs to make comparisons between the tortoise and the hare as they tell the story.

Let's revise

- Ask learners to work in groups. Tell them to list as many different places to go or things to see for a school trip as they can in a limited amount of time.
- Compare the lists from each group as a class. How many different places or attractions can the class name in total?

Top tips

- Challenge more confident learners to give synonyms for the words in the second *Practise* activity on page 14 (*greedy, clever, lovely, graceful, beautiful, stupid*). They may need to use a dictionary or thesaurus to do this.
- Extend the challenge further by asking learners to provide antonyms (opposites).

Let's go! Worksheet 3: Stories

- Worksheet 3 provides further revision about stories.
- First, learners decide if the statements about stories are true or false.
- Then they work with a partner to tell each other about a traditional story that they know.

Answers

1

Statement	True	False
Stories have characters.	✔	
We use the present simple tense for stories.		✔
Characters must be people.		✔
Stories have a start, a middle and an ending.	✔	
Some stories have a moral or message.	✔	
The setting is where a story takes place.	✔	

Worksheet 3

Stories

1 Are these sentences true or false? Put a tick (✓) in the correct column in the table.

Statement	True	False
Stories have characters.		
We use the present simple tense for stories.		
Characters must be people.		
Stories have a start, a middle and an ending.		
Some stories have a moral or message.		
The setting is where a story takes place.		

2 Work with a partner. Think of a traditional story that you know and talk to your partner about it. Do not retell the story. Tell your partner these facts:

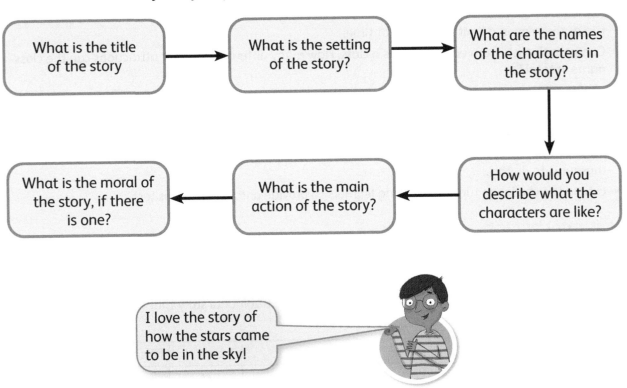

Section 4: What do you think?

 Study Guide pages 16–19

 Audio 1.8, 1.9

 Key skills: Recognising opinions of others; giving your own opinions; listening for details; using the present continuous tense; developing critical thinking skills.

Recognising opinions

Study Guide (pages 16–17)

Learn

- It is very important that learners are able to give not only their own opinions, but also to recognise and understand the opinions of others. Learners need to be able to understand the opinions they hear and to respond to them appropriately.

- The information in the *Learn* box provides some strategies learners can use to help them with recognising opinions. You could read through the strategies with the class or put learners into pairs or small groups to read about the strategies together. It may be helpful for learners to note down the key opinion phrases and opinion adverbs.

Try this

- This is a listening comprehension. It is always useful to play the audio for listening exercises at least twice. Tell learners to read the questions before playing the audio.

- Play *Audio 1.8* from the online resources at www.hoddereducation.co.uk/cambridgeextras. Learners will listen to Raj talking about books and answer multiple-choice questions about what he says. The *Hint* box encourages learners to read all of the options and to choose the best in each case, instead of guessing. There are five questions in total.

Answers

1 B   2 A   3 C   4 B   5 B

Let's talk

- Now it is learners' turn to give their opinion about types of books. Remind learners that they can use verbs and adverbs to make their opinions clearer. They should give reasons for their opinions to help other people to understand them, as explained in the *Hint* box.

- Learners should write sentences to give their opinions about the six types of books listed. Then they can use their sentences to talk about their opinions with a partner or in a small group.

Expressing opinions

Study Guide (pages 18–19)

Learn

- Learners now have opportunity to deepen and consolidate their ability to express their own opinions. Ask volunteers to read the speech bubbles about giving opinions.

- Write the key phrases for giving opinions on the board and then ask volunteers to give you an example of a sentence using each phrase to give their opinion about something. For example: *I believe poems are better than stories. I think we should recycle more. I know it is healthy to drink water, but I really like cola!*

- Encourage more confident students to respond to their classmates' sentences by agreeing or disagreeing with them. For example: *I don't believe that, stories are better because they are longer. I think you are right; recycling is very important! I disagree, cola isn't unhealthy if you don't drink it too often.*

Try this

- Before starting the *Try this* activity, take a moment to refer to the *Hint* box and revise how to use the present continuous tense in English. This consists of the present tense of the verb 'to be' and the '-ing' form of the action verb. Learners will hear this tense during the argument in the audio recording.

- This listening exercise is an audio recording of a brother and sister arguing about recycling. Play *Audio 1.9* from the online resources at www.hoddereducation.co.uk/cambridgeextras.

- Ask learners to read the questions before playing the audio recording for the first time. Then on the second listen, tell learners to choose their answers from the multiple-choice options provided.

Answers

1 B   2 B   3 A   4 B

Practise

- This writing exercise has two tasks for learners to complete. The first task consolidates understanding of the present continuous tense and the second task strengthens recognising and giving opinions.

- In question 1, learners rewrite five sentences, converting the verbs in brackets from the present simple tense to the present continuous form. In question 2, learners indicate whether they agree or disagree with five statements. When they disagree, they write a sentence in response to say what they think.

Answers

1 a What are you doing?

   b I'm washing the bottles for recycling.

   c Your four bottle are not going to make a difference.

   d Are you saying that all recycling programmes are useless?

   e I know the bottle I am recycling will be used again.

2 Learners' own responses.

Learn

- This final part of Section 4 helps to develop critical thinking skills, as well as English language skills. Opinions and facts are not the same. Facts are true statements that can be proven, while opinions are people's beliefs or feelings about things.

- Can learners tell the difference between facts and opinions? Ask volunteers to give you examples of sentences that are facts and sentences that are opinions, or prepare some sentences in advance. Say these sentences to the class and ask learners to vote whether they think each sentence is a fact or an opinion.

Let's talk

- This is an opportunity for learners to express their opinions about recycling. They are challenged to say at least five sentences about what they think, and they should use the key verbs and phrases given in bold on page 19.

- Learners plan their sentences in the space provided before sharing their opinion with a partner, in a small group or as a class.

Let's revise

- Revise the use of adverbs with positive and negative opinions. Start with the adverbs on page 16 and ask learners what other adverbs they know.

- Don't be afraid to prompt learners by questioning them in order to elicit their opinions. Ask questions such as *Why do you think that?* when learners state opinions without giving reasons for them.

- Play the two-hand challenge game.
  - Each learner must come up with five sentences that are facts, one for each digit on their left hand, and another five sentences that are opinions, one for each digit on their right hand.
  - When it is their turn to say their sentences, they start by holding their arms up with clenched fists and they unroll a digit with each sentence until, eventually, they have two palms with all digits

outstretched.

- For fun, they can shake their hands like jazz hands when they finish the challenge. Learners may be interested to know that, in sign language, shaking two hands in this way is universally acknowledged as the sign for applause (clapping).

Top tips

- You could give learners some of Roger McGough's poems to read aloud in class, to provide them more practice.
- Alternatively, provide more listening practice by allowing learners to listen to Roger McGough reading his poetry aloud – there are many videos to choose from on websites such as YouTube.

Let's go! Worksheet 4: What do you think?

- Worksheet 4 provides further reinforcement and practice in giving opinions.
- Learners differentiate between facts and opinions, and write their own sentences about recycling. Then they discuss their opinion with a partner.
- They go on to think about different types of books and discuss their opinions about them with a partner.

Answers

1

Statement	Fact	Opinion
This is an English sentence.	✔	
I think everyone should learn English.		✔
English is fun!		✔
I believe recycling is good for the planet.		✔
Paper and cardboard can be recycled.	✔	
It is okay for people to have different opinions.	✔	

2–3 Learners' own responses.

Worksheet 4

What do you think?

1   Are these sentences facts or opinions? Put a tick (✓) in the correct column in the table.

Statement	Fact	Opinion
This is an English sentence.		
I think everyone should learn English.		
English is fun!		
I believe recycling is good for the planet.		
Paper and cardboard can be recycled.		
It is okay for people to have different opinions.		

2   a   Think about recycling. Write a sentence that is a fact and a sentence that is an opinion.

   Fact: _____

   _____

   Opinion: _____

   _____

   b   Compare sentences with a partner. Talk about whether you agree or disagree with each other.

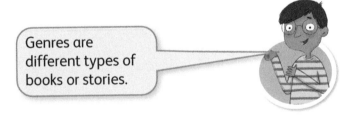

Genres are different types of books or stories.

3   a   Look at the list of genres below. Tick the ones you have read.

   • Folktales ☐

   • Science fiction and fantasy ☐

   • Adventure stories ☐

   • Comic books and manga ☐

   • Poetry ☐

   • Non-fiction ☐

   b   Compare lists with a partner. Talk about which genres you like most and least. Give reasons for your opinions.

Section 5: The short version, please!

 Study Guide pages 20–21

 Audio 1.10

 Key skills: Summarising (verbally and in writing); listening for main points; using connective words.

Key points and summaries

> **Study Guide** (pages 20–21)

Learn

- Section 5 is about describing and understanding key points and summarising information.
- Explain to learners that we need to do this sometimes to make the important information very clear, so that it isn't lost in sentences that are too long or contain too many other facts.

Try this

- This *Try this* activity is a listening comprehension. It is always useful to play the audio recording for listening exercises at least twice. Tell learners to read the information at the start of the activity and all the questions before playing the audio recording.
- Play *Audio 1.10* from the online resources at www.hoddereducation.co.uk/cambridgeextras. Kati is trying to explain to her mother where her bike is. On the first listen, as suggested in the rubric, learners could make notes on paper or in notebooks that will help them to remember things. Then, on the second listen, learners could use their notes as well as what they hear to answer questions 1–5 by choosing from the multiple-choice options provided.

Answers

1 B   **2** B   **3** C   **4** C   **5** B

Hint

- Refer to the *Hint* box to remind learners about connectives. They join parts of sentences together and help to separate clauses.
- Connectives are very useful words in English, but we must be careful not to overuse them because they can make sentences very long and complicated – as learners will see in the next activity.
- Common connectives include *and*, *so* and *but*. Which other connectives can learners recall? Can they give you examples of sentences using connectives?

Practise

- This exercise will deepen learners' understanding of sentence structure and the importance of breaking long, complex sentences into shorter, clearer sentences.
- It may be useful for learners to work with a partner first to read the text together and share ideas about how it could be broken down into smaller sentences by replacing connectives with full stops.
- Learners can then work individually to put pen to paper and make their own changes to the text. They could write their new version of the text in their notebooks for you to check.

Answers

<u>And</u> then she said why not go across it on your bike <u>and </u>that was a cool idea, <u>so</u> we made a bit of a ramp, <u>so</u> I could get up some speed <u>and </u>then I went down the ramp pedalling like mad <u>and</u> got onto the first plank <u>and</u> it held, <u>but</u> then when I was almost at the end of the plank, my weight was too much <u>and</u> the plank came up behind me <u>and</u> hit me on the head <u>and</u> then the bike hit the life jackets <u>and</u>, well, me <u>and</u> the bike went over into the water – you know that deep bit, right in the middle?

And then she said, 'Why not go across it on your bike?' That was a cool idea, so we made a bit of a ramp, so I could get up some speed. Then I went down the ramp pedalling like mad and got onto the first plank and it held. When I was almost at the end of the plank, my weight was too much and the plank came up behind me and hit me on the head! Then the bike hit the life jackets and, well, me and the bike went over into the water – you know that deep bit, right in the middle?

Learn

- Learners are familiar with listening for the main points, so this can be a good place to begin to think about summarising skills. When learners listen for main points, they are essentially picking out the points they would use to write a summary or give a verbal summary.
- Summarising is when we provide only the key information about something and we leave out any extra details. So, for example, we don't include adjectives or adverbs when we summarise; we don't include extra instructions or examples; and we don't give detailed explanations.
- Ask volunteers to read the speech bubbles.

Let's talk

- This is an opportunity for learners to practise summarising. Learners read Kati's explanation and break it down to a summary, which consists of no more than five short sentences. They should follow the steps described in the speech bubbles in the *Learn* box to do this.
- You may wish to pair up less-confident learners with more confident learners so they can work together on the summary for peer support.

Let's revise

- Revise vocabulary for things that can be recycled. How many things can learners name?
- Find the adjectives in Kati's explanation on page 21 and ask learners to provide synonyms and/or antonyms for them.

Top tips

Provide further summarising practice by asking learners to say or write a summary about what they did over the weekend – they will need to condense two days of activity into a few short sentences.

Let's go! Worksheet 5: Write a summary

- Worksheet 5 provides further reinforcement and practice in summarising as learners are asked to write a summary of the story of *The clever fox and the greedy crow*, which they may recall from Section 3.
- Learners read their summaries to a partner, and discuss the similarities and differences between their summaries.

Worksheet 5

Write a summary

Do you remember the story of *The clever fox and the greedy crow*?

1 Read the story again. Then, in the space below, write a summary of the story.

The clever fox and the greedy crow

A crow stole a piece of cheese from a window and flew up into a tall tree. A fox, who saw this happen, said to himself, 'If I am clever, I will have cheese for supper tonight.'

'Good afternoon, Miss Crow,' said Fox. 'How lovely you look today. Your feathers shine in the light. Your wings are strong and beautiful. I am sure that if you had a voice, you would sing as sweetly as a songbird.'

Crow, pleased with this praise, wanted to prove that she could sing. But, as soon as she opened her mouth, the cheese fell to the ground, and Fox snapped it up.

As he walked off he called back to Crow, 'I said a lot about your beauty, but I said nothing about your brains.'

The moral of this story is: Don't be fooled by flattery!

_____

_____

_____

_____

_____

_____

_____

_____

_____

_____

_____

2 Share and compare summaries with a partner. How are they similar and how are they different?

Section 6: Arguments

 Study Guide pages 22–24

 Audio 1.11

 Key skills: Responding to arguments; listening for details; identifying and expressing opinion; giving a short speech or talk; developing teamwork skills.

Arguing your viewpoint

Study Guide (pages 22–24)

Learn

- This final section of Unit 1 is about acknowledging the opinions of others and responding by putting forward your own point of view about the same thing. This is called arguing, which is when two or more people have different opinions about the same thing.
- Explain to learners that it is a misconception to say that an argument is always a row, with raised voices and angry people. An argument is simply listening to other people's viewpoints and maintaining your own, different opinion.

Try this

- This is a listening comprehension. It is always useful to play the audio for listening exercises at least twice. Tell learners to read the information at the start of the activity and all the questions before playing the audio.
- Play *Audio 1.11* from the online resources at www.hoddereducation.co.uk/cambridgeextras. Learners will listen to an audio recording about a customer who is trying to solve a problem with a phone she has bought, and she has called customer services for some assistance.

Answers

1 A   2 A   3 B   4 A   5 B

Practise

- This writing exercise has two tasks for learners to complete. The first task consolidates recognising when an opinion is being given and the second task allows learners to give opinions of their own.
- First, learners identify phrases in sentences which give opinions. Then they write sentences to give their own opinions about five topics.
- Tell learners that they can look back to the *Learn* box on page 22 for a reminder of useful phrases for their sentences.

Answers

1 **a** I believe   **b** I feel
  c I think   **d** I don't believe
  e In my opinion
2 Learners' own responses.

Learn

- Learners may be familiar with debating from lessons, in which case they will understand that in a debate, an argument is presented in the form of a short speech.
- To argue a point, it must be stated, then explained, and then a conclusion must be given. This process is explained in the *Learn* box and key phrases for each stage are provided.
- Ask a volunteer to read Pia's speech bubble.

Let's talk

- This is an opportunity for learners to write their own short speech to present their opinions about a statement that is made. There are three statements for learners to choose from, which reflect aspects of society that are relevant today.
- Learners plan their argument for or against the statement. Their plan should have three parts: first – stating their opinion, second – giving reasons, and third – providing a conclusion.
- When they have planned their argument, learners can present it to a partner, in a small group or to the class. Remind learners to be polite, and not to shout or raise their voices.
- When learners have given their presentations, they could reflect on their work and give each other constructive feedback about things they did well and things they could do better.

Let's talk

- This is a fun game to conclude Unit 1. There are six colourful pictures on page 24 of the Study Guide. Learners work with a partner to take turns to describe one of the items pictured for their partner to guess.
- The *Hint* box reminds learners that they can talk about shape, colour, movement or cost of the items. Learners should also consider what the items do, as well as where and how we use them.

Let's revise

- The objects on page 24 include a cricket bat and a football. Revise vocabulary for other sports and sports equipment.
- You could take some sports equipment to class and ask learners to name the items.

Top tips

- With a confident class, you could choose one of the statements from page 23 and develop it into a whole class debate.
- Learners will need to collaborate and work together to form arguments for or against the statement.
- Then hold the debate in class, ask some special guests to watch if you wish, and at the end hold a vote to decide which team presented the most convincing arguments.

Let's go! Worksheet 6: Arguing viewpoints

- Worksheet 6 provides further reinforcement and practice in arguing viewpoints. Learners look at photographs of six different items and then choose which one they think is the best. They can choose any item – there is no right or wrong choice.
- Learners should have at least three reasons for their decision and they use these to convince others that their choice is the best choice.

Worksheet 6

Arguing viewpoints

1   Look again at these items from page 24 of the Study Guide. Decide which one you think is the best and tick (✔) it. You need to think of at least three reasons for your choice. Write them down.

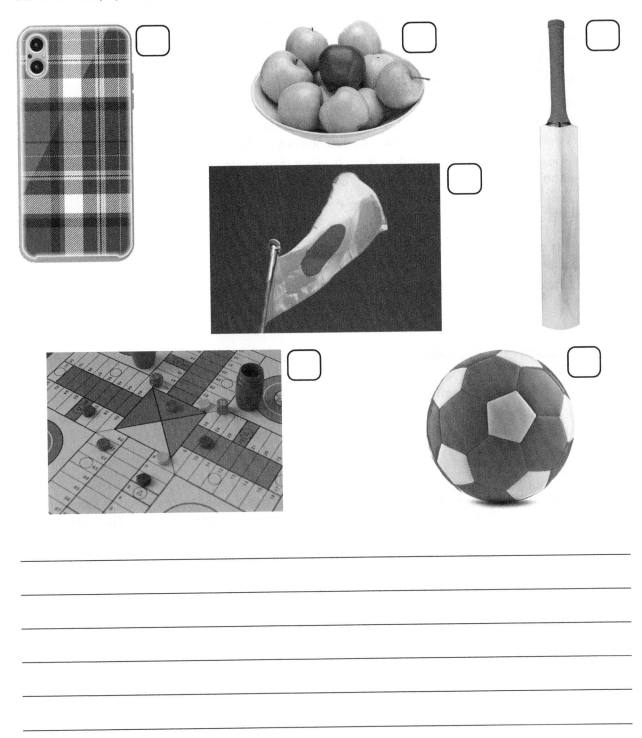

_____

_____

_____

_____

_____

_____

2   Work in a small group.
   • Try to convince others in your group why the item you have chosen is the best. Listen to their feedback to find out if they agree or disagree with you.
   • Listen to others as they explain their choices and give them feedback about why you agree or disagree with them.

Check your understanding

 Study Guide pages 25–26

 Audio 1.12, 1.13, 1.14

- Use these end-of-unit exercises to check how well learners understand the speaking and listening concepts that have been revised in Unit 1.
- The first activity on page 25 has five questions, each with three multiple choice picture options. Learners choose the best option to answer each question. Play *Audio 1.12* from the online resources at www.hoddereducation.co.uk/cambridgeextras twice and allow time for learners to make their choice, pausing the audio recording if necessary.
- The second activity on page 26 (question 1) is not multiple choice. This time, learners listen to information about a poetry competition and fill in the missing information about the rules of the competition. Play *Audio 1.13* from the online resources at www.hoddereducation.co.uk/cambridgeextras.
- The final activity on page 26 (question 2) involves listening for specific details. Allow time for learners to read all of the options for each question carefully before playing the audio recording. Then play *Audio 1.14* from the online resources at www.hoddereducation.co.uk/cambridgeextras at least twice.
- Refer learners to the *Hint* box for information about a useful strategy for these types of listening tasks.

Answers

1 a B
  b B
  c B
  d A
  e C

2 The poem must be about **water**.

  The poem must be longer than **five** lines.

  Send your entries by Friday, 14 **August**.

  Provide your name and a contact phone number or **email** address.

  You may only send us a poem that you have written **yourself**.

3 a C
  b A
  c B

Section 1: Making new words

 Study Guide pages 27–28

 Key skills: Recognising and forming regular and irregular plurals; recognising and using suffixes (word endings).

Plurals and word forms

Study Guide (pages 27–28)

Learn

- The information in the *Learn* box explains how to form plurals in English. We use plurals when there is more than one person or thing.
- There are some rules (patterns) that we can follow for regular nouns, but there are also many irregular nouns that we need to learn on an individual basis. Some nouns, such as 'sheep' are irregular because they do not change at all when we make the plural form.
- The table on the left (Rules / Examples) revises the rules for forming regular plurals and the table on the right (Noun / Irregular plural form) gives examples of some common irregular plurals.
- Ask learners to give you examples of other words that are formed by each of the rules, in addition to those shown in the table. Challenge learners to cover the irregular plurals and to see how many they can recall before uncovering the plurals.

Practise

- These questions revise and consolidate recognising and forming plural nouns. The first exercise is a word search puzzle to find 10 plural words. The *Hint* box provides a helpful strategy for learners to use.
- The second exercise requires learners to convert eight nouns from the singular form to the plural by adding the '-s' or '-es' ending.
- Encourage learners to check their answers in a dictionary to self-mark their work or allow learners to swap with a partner and mark each other's work.

Answers

1

m	o	n	k	e	y	s	c	x	g	f	l
o	x	i	v	e	r	t	e	b	r	a	e
u	o	e	i	u	k	o	g	h	a	d	p
n	p	s	l	e	t	r	e	e	s	c	u
t	e	e	t	h	j	i	d	m	s	i	p
a	z	y	b	b	s	e	s	n	e	y	p
i	i	q	d	q	b	s	a	b	s	e	i
n	m	a	t	c	h	e	s	v	h	q	e
s	k	z	b	i	c	y	c	l	e	s	s

2 **a** plants **b** buses **c** volcanoes **d** flies
   e wishes **f** nurses **g** rivers **h** bushes

Learn

- This information builds on the concept of plurals. Plurals are created by changing the ending of nouns and when we change the ending, the spelling of a word is often affected. Another word for 'ending' is 'suffix', although learners do not need to know this word.
- When we make regular plurals, we add '-es' or '-s' to the end of the word. Adding these letters can change not only the spelling of words, but the meaning too. When we add letters to the ends of words to make plurals, we know it means there is more than one of something. In this way, we can add letters to create many different parts of speech including plurals, verb forms, adjectives and adverbs. Adding the correct letters to a word is important to make sure the word 'fits' into the sentence.
- The table in the *Learn* box lists some of the most common letters added to words and gives examples.
- Refer to the *Hint* box for some spelling rules that often apply, as these will be helpful in the *Practise* activity.
- The opposite of suffix is 'prefix'. Learners don't need to know this term either. Letters (prefixes) can also be added to the beginning of words in English.

Practise

- This is a gap-fill activity. Learners look at the photograph and then complete the description of the picture.
- To fill the gaps, learners need to take the word in brackets and add letters to the end of it. Tell learners to make sure that the words 'fit' in the sentences; in other words, the sentences must make sense. To provide some additional handwriting practice for less-confident learners, you could ask them to write out the paragraph in full in their notebook or on a separate sheet of paper, using their best handwriting.

Answers

1 The city is **beautiful** tonight. The lights are **amazing**.

2 Cars and buses are **driving slowly** down the street.

3 There are people too. They are **shopping** or **looking** at shop windows. Some are with friends. They are **chatting** as they walk. Others are alone, **going** home.

4 The weather is good. It is not **cloudy** or **rainy**.

Let's revise

- Ask learners to tell you the plural words for items in the classroom and around school.
- Do some reverse plurals. Give learners a list of nouns (regular and irregular) that are already plural and ask them to tell you the singular words. Can learners apply the rules in reverse and find the original endings?

Top tips

Adverbs and adjectives are formed by adding letters at the end. Challenge learners by encouraging them to rewrite or repeat sentences and to extend them by adding appropriate adjectives or adverbs.

Let's go! Worksheet 7: Plurals and word forms

- Worksheet 7 provides further reinforcement and practice with making plurals and constructing sentences. This task can be differentiated by asking more confident learners to write, longer more complex sentences with several clauses, while less-confident learners can write shorter sentences to state the number of each item in words.
- Numbers here range from single-digit numbers to 1000. Learners might imagine sentences that could be from stories or poems, or they could be factual statements. Encourage learners to be as creative and imaginative as they can.
- When learners have finished the worksheet, share and compare some example sentences as a class. Check that learners use the correct plural for each item and that the number is spelled correctly in words.

Answers

Learners' own sentences, but including these items and quantities: *twelve eggs, three churches, eight hats, one hundred bottles, forty-four foxes, one thousand clouds, five hundred and ninety books, ten bridges, seven boats, sixty-six sheep.*

Worksheet 7

Plurals and word forms

Write a sentence in words that includes the number of each item. The first has been done for you as an example. Use your imagination to think of unusual situations!

Number	Item	Sentence
12		Dad needs 12 eggs to bake cakes for the party.
3		
8		
100		
44		
1000		
590		
10		
7		
66		

Section 2: Tricky vowels

 Study Guide pages 29–30

 Key skills: Recognising vowels and vowel sounds; using spelling rules and patterns with vowels.

Spelling rules for vowels

Study Guide (pages 29–30)

Learn

- Ask volunteers to tell you what the five vowels are (*a, e, i, o, u*).
- Ask a volunteer to tell you what word we use to describe all the other letters of the alphabet (consonants). It is important to identify the vowels in English words because the vowels dictate the sound of words; different combinations of vowels make different sounds.
- The information here is about words with long vowel sounds. Ask volunteers to read the speech bubbles, and then work through the important points to remember about spelling words with long vowel sounds.

Practise

- This is a say, copy, write and check activity for spelling words with long vowel sounds. Learners say each word aloud to practise the long vowel sounds first. Then they copy each word to practise the spelling.
- The third part of the exercise is for learners to cover the words and write them again to see how many correct spellings they can recall. It may be useful for learners to work in pairs, so that one learner can call out the words for their partner to write down, or you could call out the words to the class as a group.

Try this

This activity is about recognising correct spellings of words with long vowel sounds. There are eight words, each with three possible spellings; learners circle the one that is correct.

Answers

1 where	**2** rice	**3** boat	**4** believe
5 green	**6** rain	**7** hour	**8** toast

Learn

- Write the vowels 'e' and 'i' on the board. Remind learners that these vowels often appear together in English words. Together, they can create the long vowel sound /ee/. The tricky thing is to know the order in which they appear – is it 'ie' or 'ei'? Generally, we use the rule that we put the 'i' before the 'e', unless they appear after the letter 'c', but this is not always true.
- Ask a volunteer to read the speech bubbles for examples and exceptions to the rule.
- This vowel combination can also create other sounds in English including /eye/ and /ay/. Can learners give you any other examples of words with these sounds, other than those provided in the *Learn* box?

Practise

- In this *Practise* activity, learners see some phone screens with letters. Learners need to colour letters to find a word that has the long vowel sound /ee/. The first screen has been done as an example for learners to follow – the letters can form the word 'piece'. *Note:* Letters may or may not join in sequence, and they simply need to be coloured. They do not have to jump from one letter to the next in order.

Answers

piece, believe, niece, tree, thirty, scene

Hint

This is advice about a strategy to help with learning tricky spellings. Learners could make lists of words that they find difficult to spell and then use the *say, copy, write* and *check* method for practice.

Let's revise

- Listing is a good strategy to help learners deepen their understanding of patterns such as spelling rules. Seeing lots and lots of words with the same pattern strengthens understanding of the rule.
- For example, challenge learners to write as many words as they can that end in the letter 'y' with a vowel, or words that have particular vowel combinations. You could do this as a timed activity and then have learners share and compare their lists of words to make one longer list from the whole group.

Top tips

- To provide differentiated support for less-confident learners, spend some time categorising nouns into groups according to their sounds.
- Give learners lists of nouns to say aloud and ask them to write them down in groups. Then you can check spellings (focus on vowel combinations), as well as recognition of sound patterns.

Let's go! Worksheet 8: Spelling rules for vowels

- Worksheet 8 provides further reinforcement and practice with recognising and using spelling rules for vowels.
- First, learners identify whether 12 words have the correct spelling. Then they rewrite the words they think are incorrect. Finally, they choose any two words form the original list to use in sentences.
- This could be differentiated by asking more confident learners to produce more than two sentences, or by asking them to use more than one of the words in a single sentence.

Answers

Word	Correct (✔)	Incorrect (✗)	Correct spelling
science	✔		
thier		✗	their
there	✔		
wear	✔		
acheive		✗	achieve
pleese		✗	please
where	✔		
eight	✔		
cieling		✗	ceiling
weird	✔		
meel		✗	meal
bloo		✗	blue

Worksheet 8

Spelling rules for vowels

1 Complete the table. Put a tick (✔) if you think the spelling is correct and put a cross (✗) if you think the spelling is incorrect.

Word	Correct (✔)	Incorrect (✗)
science		
thier		
there		
wear		
acheive		
pleese		
where		
eight		
cieling		
weird		
meel		
bloo		

2 Rewrite the words that you think are incorrect, spelling them correctly.

Word	Correct spelling
science	
thier	
there	
wear	
acheive	
pleese	
where	
eight	
cieling	
weird	
meel	
bloo	

3 Choose at least two of the words from the table. Use them in sentences.

_____

_____

_____

_____

Section 3: Tricky consonants

 Study Guide pages 31–32

 Key skills: Recognising consonant combinations and their sounds sounds; using spelling rules and patterns with consonants.

Spelling rules for consonants

Study Guide (pages 31–32)

Learn

- This section moves away from vowels and looks at consonant combinations, in particular the combinations of 'c', 'k' and 'h'. Explore the information in the *Learn* box to help learners recall when to use 'c' and when to use 'k', as there are some rules we can follow.
- Words that are spelled with the consonant combinations '-ch' or '-ck' need to be learned individually, there are no set rules.
- Read the examples from the table and ask learners to repeat them back to you. Focus on the sound differences of words spelled with a 'c' – sometimes 'c' sounds like /s/ and sometimes it sounds like /kuh/.
- Ask learners to give you more example of their own to add to each group of words in the table.

Hint

This useful tip reminds learners that they can use a dictionary to check tricky spellings. You could have dictionaries available in the classroom for this.

Practise

- This *Practise* activity is about recognising incorrect spellings. Learners are given ten pairs of words, but only one word is correct in each pair. They need to identify the correct spellings and circle them.
- Encourage learners to say the words aloud and to listen to the sounds to help them make their choice if they are unsure.

Answers

1 crayon	**2** climb
3 computer	**4** kitten
5 kick	**6** cucumber
7 pocket	**8** clock
9 chameleon	**10** ache

Try this

- Tongue twisters are short sentences that are tricky to say, and very tricky to say quickly. This is because they have repeated sounds or several sounds that are very similar, such as the same initial sound on all words in the sentence. The example sentence has repetition of the /kuh/ sound.
- Challenge the class to read the example tongue twister aloud together. Do this several times, getting faster and faster each time.
- Learners then write their own tongue twisters. Tell them to look at the advice in the *Hint* box about how to structure their tongue twister. If they are struggling to think of words with the /kuh/ sound, tell them to look back to the words on page 31 in the Study Guide for inspiration.
- When they have finished writing, ask learners to read over their work to check spellings and punctuation. Learners could swap with a partner and challenge them to say their tongue twister.

Learn

Silent letters are letters that are always included when we write words but which are never included when we say words. Explain to learners that we do not say the sounds of silent letters, which can make them tricky to spell correctly.

Try this

Do this activity as a class for consolidation. Ask volunteers to say each word aloud and to identify the silent letters. The silent letters are shown in bold in the answers below.

Answers

1 pip**e**

2 clim**b**

3 bri**d**ge

4 **k**nife

5 **w**rite

Practise

- Learners see seven more words that have silent letters and identify the silent letters by circling them.
- Encourage learners to say the words and to listen to the sounds to help them to find the silent letters. The silent letters are shown in bold in the answers below.

Answers

1 **k**nob   2 **w**rote

3 ey**e**   4 **w**hole

5 g**u**est   6 san**d**wich

7 thum**b**

Learn

- Learners have seen the consonant combinations '-ck' and '-ch' already in this section, but these are not the only consonant combinations we can find in English words. Sometimes, two of the same consonants can appear together (this is called a double consonant). In fact, sometimes we must have a double consonant as a rule when we add letters (a suffix) to the end of a word.
- Ask a volunteer to read the speech bubbles for examples of words that have a double consonant or change to have a double consonant. Challenge learners to give you more examples of words they know that have a double consonant – this could appear within a word or in a suffix.

Practise

- This final *Practise* activity provides consolidation of English words with double consonants.
- Ask learners to use neat handwriting when they copy the words. Then they need to think of more words for each double consonant and make a list.
- You could practise the spellings again by having a written spelling test or a verbal spelling bee, where learners spell each word aloud letter by letter.

Let's revise

- Give learners a list of the letters of the alphabet with the vowels removed. Challenge them to write down one word, any word, for each letter. The word should contain that letter somewhere.
- To differentiate this for more confident learners, tell them that the letter cannot be at the start of the word, for example for the letter 'd', they could say 'hand' but not 'dog'.

Top tips

Tongue twisters can be lots of fun and there are many well-known English tongue twisters you could try chanting with the class. See if learners can say each tongue twister slowly, and then repeat it, getting quicker and quicker each time. Some tongue twisters you could use are:

- Red lorry, yellow lorry, red lorry, yellow lorry.
- She sells seashells on the seashore.
- Fred fed Ted bread, and Ted fed Fred bread.

Let's go! Worksheet 9: Spelling rules for consonants

- Worksheet 9 provides further reinforcement and practice with recognising and using spelling rules for consonants.
- First, learners identify whether 12 words have the correct spelling. Then they rewrite the words they think are incorrect. Finally, they choose any two words from the original list to use in sentences.
- This could be differentiated by asking more confident learners to produce more than two sentences, or by asking them to use more than one of the words in a single sentence.

Answers

1–2

Word	Correct (✔)	Incorrect (✗)	Correct spelling
roket		✗	rocket
choir	✔		
kik		✗	kick
packet	✔		
fizy		✗	fizzy
trafick		✗	traffic
knife	✔		
lory		✗	lorry
whole	✔		
chicken	✔		
carott		✗	carrot
hopping	✔		

3 Learners' own responses.

Worksheet 9

Spelling rules for consonants

1 Complete the table. Put a tick (✓) if you think the spelling is correct and put a cross (✗) if you think the spelling is incorrect.

Word	Correct (✔)	Incorrect (✗)
roket		
choir		
kik		
packet		
fizy		
trafick		
knife		
lory		
whole		
chicken		
carott		
hopping		

2 Rewrite the words that you think are incorrect, spelling them correctly.

Word	Correct spelling
roket	
choir	
kik	
packet	
fizy	
trafick	
knife	
lory	
whole	
chicken	
carott	
hopping	

3 Choose at least two of the words from the table. Use them in sentences.

_____

_____

_____

_____

Section 4: Don't get confused!

 Study Guide pages 33–34

 Key skills: Using the apostrophe corectly; recognising English homophones.

Apostrophes and homophones

Study Guide (pages 33–34)

Learn

- Contractions are short forms of words in English. They are formed by removing a letter or letters from a word, or by combining two words together and using an apostrophe to show where the letters have been removed. For example, 'who is' becomes 'who's' and 'it is' becomes 'it's'.
- Ask volunteers to tell you examples of other contractions that they know and write them on the board.
- The apostrophe is also used in English to show possession by adding apostrophe + 's' to the end of nouns. For example, 'Paul's book' tells us that the book belongs to Paul.

Practise

- In this activity, learners read six sentences. In each case, there is a choice of two words, and learners need to decide which word correctly fits into the sentence.
- Ask learners to read the reminder in the speech bubble before they begin. Less-confident learners could work with a partner for peer support.
- Extend this activity by asking learners if they have ever tried skateboarding, running or yoga, and encourage them to share their experiences. Ask them to talk about when they did the activity, if they used any special equipment, if they enjoyed it and if they would like to do it again.

Answers

1 We're	2 Who's	3 He's
4 it's	5 your	6 who's

Try this

This activity requires learners to complete sentences that form part of a dialogue between three people. Learners could do this individually and then share and compare their dialogues, or they could work in a group of three to construct the sentences together and then perform their dialogue for the class.

Learn

- Homophones are words that sound the same but are spelled differently and can have different meanings, for example 'to', 'two' and 'too'.
- Ask volunteers to read the speech bubbles aloud.

Try this

- Learners see four pairs of homophones and are challenged to write a sentence for each word. To do this, it is crucial that learners know and understand the different meanings of each word in each pair of homophones. Ask learners what they think they should do if they are not sure.
- Have dictionaries on hand for learners to use to look up meanings of words. When they have finished, learners can swap with a partner to share and compare each other's sentences.

Hint

- This advice in the *Hint* box is about taking care with words that have very similar spellings, such as 'of' and 'off'. These are not homophones but are often easily confused by ESL learners, so it is good practice to be able to use them correctly.
- You could ask more confident learners to say or write sentences using the words in the *Hint* box.

Let's revise

- Give learners lists of contractions with the apostrophe missing. Challenge them to add the apostrophe to each word in the correct place.
- Give learners sentences with missing words and ask learners to choose between homophones to complete the sentences.
- Tell learners to use mnemonics to help learn tricky spellings.

Top tips

- Homophones can be used in humour, especially in jokes where the incorrect word is said deliberately!
- This short video explains one such joke, and has some helpful advice about homophones: www.youtube.com/watch?v=mNL-O7cJTMs
  Watch this video with the class or advise learners to watch it in their own time or for homework.
- Ask learners if they know any other homophone jokes, or could they make one up of their own.

Let's go! Worksheet 10: Apostrophes and homophones

- Worksheet 10 further revises the use of the apostrophe in English sentences and the recognition of homophones.
- Learners read a passage about a school trip containing deliberate mistakes involving missing apostrophes and the incorrect use of homophones. They need to rewrite the passage and make the corrections. The corrections are shown in bold in the answer below.

Answer

Class Six were talking about the class trip.

'**Who's** excited?' asked Miss Hall, **their** teacher. 'Would you like to know more about it?'

'Yes, please!' chirped the class.

'First, **we're** leaving school at 8 a.m. , so **don't** be late or **you'll** miss the bus!

Next, you all need to **wear** a jacket in case it is raining and if you want, you can bring a hat **too**. **Don't** forget your sandwiches for lunch!

Mr Jones will meet us at the museum. **He's** the person who will show us around. **He's** very clever and can answer all **your** questions about the exhibition.'

Worksheet 10

Apostrophes and homophones

1   Read this passage about a school trip.

Class Six were talking about the class trip.

'Whos excited?' asked Miss Hall, there teacher. 'Would you like to know more about it?'

'Yes please!' chirped the class.

'First, were leaving school at 8 a.m., so dont be late or youll miss the bus!

Next, you all need to where a jacket in case it is raining and if you want, you can bring a hat two. Dont forget your sandwiches for lunch!

Mr Jones will meet us at the museum. Hes the person who will show us around. Hes very clever and can answer all youre questions about the exhibition.'

2   Rewrite the passage. Fill in the missing apostrophes and look for any homophones that are incorrect. Change these homophones to the correct words for this situation.

_____

_____

_____

_____

_____

_____

_____

_____

_____

_____

_____

_____

_____

_____

Section 5: Proofreading

 Study Guide pages 35–37

 Key skills: Checking and reflecting skills; reading for details.

Postcards

Study Guide (pages 35–37)

Learn

• Section 5 focuses on the importance of learners proofreading their own work. Learners may think that proofreading is just about checking spelling, but it isn't. Proofreading includes checking that sentence structures work, that punctuation is correct, and making sure that the verbs work, as well as checking spelling.

• Proofreading one's work means that it will always be the best it can be. It is important for learners to take responsibility for proofreading their work so that it becomes a natural stage of the process whenever they produce written work.

• Read through the proofreading list in the *Learn* box and ask learners how many of them do these things when they are doing written work.

Try this

• This is a reading comprehension. Tell learners to read the postcard on page 35 to themselves. Then put learners into pairs and tell them to talk about the mistakes they have found in the postcard.

• Then share and compare ideas together as a class. What reasons can learners suggest for the mistakes that have been made? Has Janesh missed consonant combinations, or made the wrong choice of homophones, for example? Remind learners that many of Janesh's mistakes could have been avoided by proofreading.

• After talking through the errors, learners can complete question 2 of the activity, which involves identifying the mistakes and then writing out a correct, edited version of the postcard.

• The answer below is a suggestion, with a minimum amount of rewriting and correcting. More confident learners may do significantly more rewriting, for example by swapping 'incredible' and 'amazing' for other words to avoid repetition.

Answer

Learners' own responses, for example:

Hi Ratty!

You will not believe this, bro! We had the most amazing, fantastic, incredible holiday!

We went on an awesome white-water ride down the Ganga River. Amazing! What an incredible ride! We all got fantastically soaking wet. I was incredibly terrified!

The water was amazing, incredibly green and blue.

Wish you could have been there! You would have had an amazing time too!

Cheers, your friend

Janesh

Practise

• In this *Practise* activity, learners write a postcard to reply to Janesh. They should write two short paragraphs about what they did on holiday. They can use their imagination to make something up if they have not had a holiday recently or can't remember.

• The annotated postcard gives learners advice about how to structure the postcard. Remind them that postcards are short with concise sentences.

41

Let's revise

- Revise vocabulary for types of holidays and holiday activities. Construct mind maps on the board and leave them there for learners to refer to.
- You could extend this further to include types of transport or clothing, or equipment or items that we take on holiday.

Top tips

When learners have written their own postcards, ensure that they proofread their work and give them an opportunity to identify and correct any mistakes. Then ask learners to swap postcards with a partner to proofread each other's work.

Let's go! Worksheet 11: Write a postcard

- Worksheet 11 provides further revision of using the correct layout to write a postcard. Learners can use the tips on page 37 of the Study Guide to help them use each part of the postcard correctly.
- Learners write about an amazing holiday or place they have visited and they also have chance to use their art and design skills by creating a picture for the front of the postcard.

Worksheet 11

Write a postcard

Write a postcard to a friend or family member about an amazing holiday or place you have visited. Remember to write on the left side and put an address on the right side. You can draw and colour a picture for the front of the postcard too!

Section 6: Punctuation

 Study Guide pages 38–40

 Key skills: Recognising punctuation; interpreting punctuation; using punctuation; using connective words; forming paragraphs from sentences.

Different punctuation marks

Study Guide (pages 38–40)

Learn

- Punctuation is an essential part of language. Without punctuation, sentences would be difficult to decipher, and the meaning of written English would not be clear.
- Ask learners to tell you examples of punctuation that they know and to say or write an example of that punctuation being used in a sentence.
- Go through the functions of the punctuation in the *Learn* box, allowing time for learners to ask questions if they are unsure about anything.

Try this

- This matching activity consolidates understanding of the functions of different types of punctuation.
- Learners draw lines from sentences to match them to the functions of the type of punctuation used in the sentence. Then they write sentences of their own using each type of punctuation.
- Remind learners to use neat handwriting for their sentences.

Answers

1 a For an exclamation
   b For a question
   c To mark contractions
   d For proper nouns
   e For the pronoun I
   f To separate items in a list
   g To mark speech
   h To mark a pause in a sentence; for proper nouns
   i For proper nouns; to mark possession

2 Learners' own responses.

Practise

This writing exercise has two tasks for learners to complete. The first task consolidates recognising when punctuation should be used as learners read an unpunctuated text and insert the punctuation. Then they write out a correct, clean version of the same paragraph.

Answer

I am Alex Rider and I'm a spy. That's my job and it runs in my family, because my uncle was a spy too. I became a spy by accident when I tried to find out more about where my uncle worked after he died in a car crash. I visited the Royal and General Bank, but his office was locked.

Practise

- Connectives are words that we use to join parts of sentences together, which means we can give more information in one sentence instead of having to use two or more shorter sentences.
- In this *Practise* activity, learners rewrite sentences and insert the connective in the correct place. Then, they use connectives to join together a group of sentences to form a paragraph, which they then write out neatly.

Answers

1 a I found a way in through a window and I discovered a file called Stormbreaker.

b The contents of the file were shocking, but I could see they were true too.

c I had to find and stop the villain before he set the plan in action.

d First, I needed to find where he was hiding, then I could make a plan.

e Suddenly, 'Help!' someone shouted. 'He's running away and he's got my bag!'

2 **Although** I'm only 15 years old, I'm a fully trained spy.

I'm sure you think I'm joking, **but** I'm not.

If you want a mystery solved, **then** I'm the spy you need.

Let's revise

Give learners a list of ten sentences of your choice, perhaps relating to a particular theme or topic. Ask learners to decide if the sentences are questions, statements or exclamations, and to insert the correct punctuation mark at the end of each sentence.

Top tips

- Take a short paragraph from a children's story, type out the paragraph and remove the punctuation. Then ask learners to rewrite the paragraph and to include the punctuation.

- The story could then be used for a group reading activity or for independent reading by more confident learners.

Let's go! Worksheet 12: Punctuation

- Worksheet 12 provides further revision of punctuation. First, learners draw punctuation marks. Then they explain what some punctuation marks are used for. Finally, they use their grammatical knowledge and critical thinking skills to weigh up the similarities and differences between apostrophes and commas.

- Ask learners to work in pairs or small groups to share their ideas for the last part of the task.

Answers

1

Full stop	.
Exclamation mark	!
Comma	,
Colon	:
Question mark	?
Apostrophe	' '
Speech marks	" "

2 Speech marks show the start and end of a quote.

Exclamation marks indicate the end of a sentence that is an exclamation, such as a shock or surprise.

Full stops indicate the end of sentences that are statements.

3 As a guide, learners should note that the punctuation marks are the same, but they are placed differently, for example the comma is always level with the line of text, but apostrophes 'float' between two letters. Commas are used to separate items in lists, and to indicate pauses in sentences. Apostrophes show contractions and can indicate possession.

Worksheet 12

Punctuation

1 Draw the punctuation marks.

Full stop	
Exclamation mark	
Comma	
Colon	
Question mark	
Apostrophe	
Speech marks	

2 Write a sentence in your own words to explain the purpose of each punctuation mark.

Punctuation	Purpose
Speech marks	
Exclamation mark	
Full stop	

3 Read the speech bubbles.

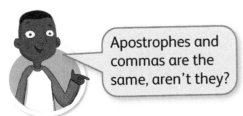

Apostrophes and commas are the same, aren't they?

No. They look the same, but they are used differently.

a How are commas and apostrophes similar, and how are they different? Explain in your own words, with examples to illustrate your reasons.

_____

_____

_____

_____

_____

_____

b Share your ideas with a partner or in a small group.

Section 7: Writing layouts

Study Guide pages 41–52

Key skills: Producing written work in a variety of formats: reports, presentations, stories, invitations, letters and emails.

Paragraphs

Study Guide (pages 41–42)

Learn

- In this section, learners will explore a variety of different types of layouts for written English. One thing that most of these layouts have in common is that they use paragraphs. So, regardless of whatever format of written work learners are doing, they need to be able to understand and use paragraphs to structure it.
- Use the information in the *Learn* box to revise the features of paragraphs and allow time for learners to ask questions if they are not sure about anything.
- Ask a volunteer to read the important rule explained in the speech bubble.

Try this

- This activity consolidates understanding of the structure of paragraphs in English.
- Learners read two paragraphs, and, in each case, they identify the topic sentence (the main idea), and then identify and count the supporting sentences.
- Learners could work on this activity individually and then compare results with a partner, or they could work together in pairs for peer support.

Practise

- In this writing activity, learners need to use their imagination and creative thinking skills. Learners see two different pictures and use them as inspiration to write a paragraph about each picture. Picture A shows a fresh loaf of bread hot from the oven; we know that it is hot because the cook is carrying it carefully with a tea towel. Picture B shows a path through a forest lined by tall trees.
- For differentiated support, you could ask less-confident learners to choose one picture to write about, while more confident learners could write about both pictures.
- If learners are struggling with what to put in their paragraph, or where to start, encourage them to think about their senses and to consider what they could see, hear, smell, feel and taste.
- Tell learners that paragraphs can be fiction or non-fiction. In other words, the bread picture could inspire a paragraph about cooking and/or staying safe in the kitchen (non-fiction), or it could inspire a paragraph about a wonderful picnic with sandwiches in a faraway land (fiction). The woodland picture could inspire a paragraph about animals or trees of the forest (non-fiction), or it could inspire a paragraph about a thrilling forest adventure (fiction).
- Extend the activity by putting learners into pairs or small groups to share and compare their paragraphs by reading them aloud. How are they similar and how are they different? Ask learners to identify the topic sentence and the supporting sentences in each other's paragraphs.

Reports and presentations

Study Guide (pages 42–43)

Learn

- The first type of layout that learners revise is for reports or presentations. Reports or presentations are non-fiction because they are about something true (factual). Reports begin with an introduction, contain details in the body section and finish with a conclusion.

- As with all types of writing, reports and presentations should be planned first. A good plan will make sure that learners don't leave out anything important and that their work has a clearly defined structure.

Try this

- This activity helps learners to identify the main features of a report about natural resources.
- Learners should read the report from start to end first, and then complete the activity to identify and label the different parts.
- Ask learners where they think we might read a report like this one.

Practise

- This is an opportunity for learners to write a report of their own. There are three options for learners to choose from as the topic for their report.
- Remind learners that they can use the example from the *Try this* activity as a guide and that they may also find some useful vocabulary there that they can use.
- Remind learners about the importance of planning. Provide some rough paper for plans, or learners could make plans in their notebooks.
- Extend the activity by putting learners into pairs or small groups to share and compare their reports by reading them aloud. How are they similar and how are they different? For learners who chose the same options, what different ideas did they have? For learners who chose different topics, why did they make those choices?
- You may also wish to use this activity for cross-curricular purposes and develop ICT skills by asking learners to type their reports on the computer; they can add pictures or frames to their work and then print them to be used in a classroom display.

Stories

Study Guide (pages 44–45)

Learn

- Stories are a type of fiction writing. They can be based on true events for inspiration, but stories are created or thought of by the writer. Writing a story is what we call creative writing, as we need to use our creativity and imagination.
- Use the information in the *Learn* box to talk about the main features of stories: the setting, the characters, the action and the ending.

Try this

- Read the story of *Red and the Wolf* with the class. This is a retelling of the well-known fairy tale *Little Red Riding Hood*. Volunteers could take turns to read aloud in class, or you may wish to put learners into groups so that more learners have the opportunity to read.
- After reading, learners should make notes about the different parts of the story. They could do this individually, in pairs or groups, or you could do this activity verbally with the whole class as a discussion.

Practise

- Learners write their own version on a well-known traditional tale. You may want to spend some time talking about traditional tales before starting, so that learners have some ideas about the types of stories they could write.
- There is space for learners to first make some brief notes as the basis of a plan and then to write their story neatly. When they have finished, ask for volunteers to read their story aloud to the class. Invite volunteers to the front and encourage them to think about emphasis and intonation when they are reading.

Invitations

> ## Study Guide (pages 46–47)

Learn

- Invitations are requests to do something. They are factual so they need to contain all the essential information that people need to know. They are not written in paragraphs, or even in complete sentences; instead, they simply have the information stated in a clear way, usually on separate lines.

- Ask learners to describe the kinds of information they think are essential for an invitation. Look for ideas such as: invitations need to include the date and time, the location, what the event is, what to wear or bring, and how to accept the invite.

- Ask volunteers to read the speech bubbles. Note that support may be needed with the third speech bubble, which explains the acronym RSVP.

Try this

- Before learners start this activity, tell them to read the advice in the *Hint* box for a reminder about useful information to include on an invitation.

- Learners read an invitation, extract the important information and make notes about it.

Practise

- This fun activity allows learners to be very imaginative as they write invitations for their own birthday party. Tell learners to imagine that they are planning their own perfect party, which can have a theme and can take place in any location they choose.

- Learners can use the invitation from the *Try this* activity for inspiration. They can also add some pictures around the invitation that are related to the event.

Letters

> ## Study Guide (pages 48–49)

Do you remember?

- This information reminds learners what letters are. Explain to learners that letters can be formal or informal. The example letter shows an informal, friendly letter.

- Letters are handwritten and sent in the post or delivered; they are not sent electronically. Ask volunteers to explain when we should use the past tense and when we should use the future tense in English, and to read the speech bubbles as examples of sentences in the past and future tenses.

Example

- This is an annotated example of a friendly letter. The main features including the address, the opening line, paragraphs and the ending are labelled.

- Put learners into small groups to read the letter and to talk about the structure of it together.

Try this

- Learners now have the opportunity to write their own informal, friendly, letter. There are three options for topics to write about in their letter, which range from holidays, to making money and using social media.

- Remind learners to use clear, neat handwriting, to structure the letter using paragraphs and to proofread their letter when they have finished.

Emails

Study Guide (pages 50–52)

Learn

- Like letters, emails can be formal or informal. Emails are typed on the computer (or on a phone, tablet or other device) and they are sent using the internet, which means they arrive immediately. People do not need to wait for emails to be delivered like letters that arrive in the post.

- Ask volunteers to read the speech bubbles. As a class, discuss what is similar and what is different about letters and emails.

Example

This annotated diagram shows all the different components of an email, including the address lines and subject bar, the meanings of Cc and Bcc, and the space for composing the email body.

Try this

- For the first part of this activity (question 1), learners use the letter they wrote on page 49 and convert it to an informal, friendly email. Learners can make up email addresses for themselves and their friend; the simplest way to do this is for them to add their names before '@email.com' (use the relevant country code for your setting to make them more authentic).

- For the second part of this activity (question 2), learners write a more formal email to apply for a job as an elephant waterer! It will be useful for learners to make a plan first on rough paper or in their notebooks. You may also wish to go over more formal ways to open and close letters or emails. Ask: *What can we do when we don't know the name of the person we are sending the email to? What greetings can we say to a person in authority, or in a formal setting?* (Use 'To whom it may concern' or 'Dear Sir/Madam'). In their email, learners should explain why they are applying for the job and why they would be good at it. Tell learners to refer to the *Hint* box for some advice about questions they could ask in their email as well.

Let's revise

- There are many options for vocabulary drills alongside the material in this section. You can drill vocabulary by creating mind maps on the board or by playing a ball game. For the ball game, learners sit or stand in a circle and pass a ball around from learner to learner, calling out an appropriate English word as they do so. The challenge of ball games can increase if they are played against the clock.

- Themes and topics that could be practised in this section include foods, cooking and kitchen equipment, cooking verbs; animals and plants / nature; the environment and resources; types of stories; or celebrations and events.

Top tips

- As a summative activity, ask learners to compare the various purposes and features of the different writing layouts they have revised in this chapter. Ask: *What features to they have in common and how are they different?* Elicit opinions, ask which type of writing they prefer and encourage them to give reasons why.

- After learners have completed the story planning task from question 4 on page 53, they could then write their stories in full in class or for homework.

Let's go! Worksheet 13: Writing

Worksheet 13 provides learners with further reinforcement and practice in different writing tasks.

Worksheet 13

Writing

Choose one of the following tasks and use the space on page 51 for your writing. As an extra challenge, complete more than one writing task. Write any additional tasks on separate sheets of paper or in your notebook.

Option A

Look at the pictures. Write a paragraph for each picture. Make sure that you have a topic sentence and several supporting sentences in your paragraphs.

Picture 1

Picture 2

Option B

Write a report about one of these topics:

- Healthy eating
- Keeping fit
- A book report for the last book that you read

Option C

Write a letter to a friend who is coming to visit you for a week during the holidays. Explain what you will be doing and tell them any plans or arrangements you have made. Think about activities to do and places to go.

Option D

Design an invitation for a birthday party. The party can have a theme of your choice if you want it to. Remember to include all of the important information that people attending the party need to know.

Worksheet 13 (continued)

Check your understanding

 Study Guide pages 53–54

- Use these end of unit exercises to check how well learners understand the language concepts that have been covered throughout Unit 2.
- The first activity (question 1) has six questions, each with words in bold that have deliberate mistakes. Learners rewrite the sentences, correcting the mistakes that are related to either spelling, or punctuation, or both.
- The second activity (question 2) requires learners to write sentences about the structure of paragraphs and this is followed by a third task (question 3) to write a paragraph about how to use classroom items. Learners go on to plan a story (question 4) but note that this is just the plan – they do not actually write the story; the assessment here is on their ability to plan out the main features for a story.
- Finally, learners compose a friendly email about surviving an adventure (question 5). This could be based on true experiences or learners could use their creative thinking skills to make up an adventure if they would prefer not to share their own experiences.

Answers

1 a who's, she's
   b off, his
   c quite, to
   d There, sitting
   e wear, to
   f We're, tomorrow
2–5 Learners' own responses.

Unit 3 Reading

Section 1: Reading for the main points

 Study Guide pages 55–56

 Key skills: Skimming and scanning; identifying main points.

Newspaper reports

Study Guide (pages 55–56)

Learn

- Learners may think that reading is simply reading, but this unit explores how there are in fact different ways to read, and different things to look out for and think about when we are reading.
- Section 1 of this unit is about reading to understand the main points, or the gist, of newspaper reports. You could take some newspapers to class to use as a conversation starter (check that the content is suitable first). English language newspaper reports can be printed out from the internet.
- Ask learners: *What do newspapers do? Are newspapers useful? What type of information do newspapers contain?*
- Explain the general structure of newspaper reports using the information in the *Learn* box. Remind learners about the importance of paragraphs and revise how paragraphs are structured.

Try this

- This activity comprises several tasks relating to a newspaper report about a volcano. First, learners read the opening paragraph and identify the main points. Then they answer comprehension questions and write their own caption for a second photograph to accompany the article.
- Some technical vocabulary is explained in the glossary box. Remind learners that they can use dictionaries to look up any other words they are not sure about.
- Note that newspaper reports are written in the past tense because they describe events that have happened. Ensure that learners understand this by asking them to give examples of verbs in the past tense from the article.

Answers

1 A volcano erupted; La Palma; months; a tropical paradise with a moonscape.

2 Learners read the newspaper report.

3 C

4 C

5 A

6 After three months

7 19 September 2021

8 Learners' own responses. Answers will vary. For guidance, learners should include destroyed buildings and homes, lava, ash and destroyed trees / plants.

Let's revise

- Revise the structural features of newspaper reports – headlines, paragraphs, quotes, pictures and captions. Ask learners to identify these features.
- Revise vocabulary for amazing places on the Earth such as volcanoes, mountains, rainforests and rivers.

Top tips

- To bring some geographical skills into the lesson, learners could use a world map, atlas or globe to locate all of the places mentioned in the newspaper report.

- To help to develop critical thinking skills, you could talk to the class about the concept of fake news. Ask: *Can we believe everything we read in newspapers?*

Let's go! Worksheet 14: Identify main points

- Worksheet 14 provides further reinforcement and practice identifying the main points.

- Learners read an article and answer multiple choice questions about it. You could do this as a quick fire or timed exercise to help to develop speed and efficiency when skimming or scanning texts for the main points. Note STEM = Science, Technology, Engineering, Maths.

Answers

| 1 C | 2 B | 3 A | 4 B | 5 B | 6 A | 7 C | 8 B |

Worksheet 14

Identify main points

PROFESSOR PRINCESS!

Crowds flocked to the town centre this afternoon for the arrival of Princess Tara who was officially opening the new Science Museum. The state-of-the-art museum has taken four years to construct and has cost upwards of $4 billion.

The princess looked excited as she arrived at the event wearing a pink dress with green accessories and waved to the crowd. She was accompanied by her cousin, Lady Samantha, who dressed in a blue coat for the chilly spring weather. Both ladies stopped to pose for photographers outside the museum.

When asked about the museum, Princess Tara said, 'This museum will be an excellent tool for teaching and education. I hope it will encourage more young people to enjoy and engage with STEM subjects and hopefully work in those industries in future!'

The princess was given a guided tour by Roger Hatfield, the general manager of the museum. She tried out some interesting experiments and had a go at the hands-on exhibitions including making her hair stand on end with static electricity.

The museum will be open to the public from tomorrow, 12 July and there are discounted tickets available for school groups. Check the museum website and email them directly for more information about exhibitions and events.

Answer the questions. Circle the correct option: **A**, **B** or **C**.

1 What type of building was opened?

   A A school   **B** An university   **C** A museum

2 Why did Samantha wear a coat?

   A It was warm.   **B** It was cool.   **C** She likes coats.

3 What does Tara do when she arrives at the building?

   A Waves at people   **B** Sings for people   **C** Ignores people

4 According to Tara, who or what is going to benefit?

   A The government   **B** Young people   **C** Roger Hatfield

5 Who is Roger Hatfield?

   A A prince   **B** The manager   **C** Tara's father

6 What happened to Tara's hair?

   A It stood on end.   **B** It got wet.   **C** It went curly.

7 What can schools get?

   A Free tickets   **B** Big tickets   **C** Cheaper tickets

8 How do people find out more information?

   A By phone   **B** By email   **C** By post

Section 2: Reading for meaning

 Study Guide pages 57–63

 Key skills: Reading at length, following a plot and understanding meaning, sentence structures and paragraphs; reading cartoon strips, and direct and reported speech.

Stories

Study Guide (pages 57–61)

Learn

- In Unit 2, learners explored how to write short stories. Now, they will read stories. Explain to learners that when we read a story, it is important that we understand the meaning, so that we don't miss any information that is important in the plot, or any descriptions that help us to imagine the scenes.

- Look at the picture of the family on a bike ride. Ask learners to imagine that this picture came from a story. What might be happening in the story? Who would the characters be? What problem might they find? How might the problem be solved? Remind learners that while pictures are very important and can help us to see what is happening, the words are even more important because when we read the words, we create our own pictures in our mind.

Try this

- This activity is about a short story called *An unusual summer holiday*. Learners should read the story individually, all the way through first. Less-confident learners could work with a partner to read the text together for support.

- First, learners work with a partner to look at the accompanying picture. They discuss what they think the story is about and who the characters are. Then they go on to answer ten questions based on the text. Responses include lifting direct quotes from the text and explaining how they feel after reading certain lines of text.

Answers

1 Learners' own responses, for example: *The story is about doing something unusual during a summer holiday. The main characters are a brother and sister (Janesh and Sitara) and they are travelling in a car with their parents. Janesh is looking out of the car window and Sitara is reading a book.*

2 It was near Rishikesh and it was fun.

3 'pestered'

4 'shaking his leg up and down in excitement'

5 'watched the countryside **flash** by' / 'we **screeched** to a halt'

6 Ganga River and Ganges River / Learners' own responses, for example: An English name (because India used to be a British colony) and an Indian name.

7 'Water wasn't his friend.' / 'Janesh was looking about the same colour as the river.'

8 Janesh was nervous / worried / feeling sick. (The water was green. In English, when we say a person looks 'green' it means that they are feeling bilious or sickly. This is often associated with being nervous about something that is going to happen.)

9 **b** – excited

10 **a** 'thrill'　　　　　　**b** 'stillness'

11 Learners' own responses. Answers will vary. Learners should describe something that could reasonably happen next. The story ends when they are approaching a rapid, so they could describe what happens then, or what happens after they have finished paddling.

Try this

- Ask volunteers to read the speech bubbles on page 59. Discuss what learners expect to happen in the story, judging by the title alone.

- Learners then read an extract from *The Selfish Giant*.
- Tell learners that the extract they will read is from the middle of the story, and then put them into small groups to read the text together. They should take turns to each read part of the text, such as a sentence at a time.
- In their groups, learners can discuss the questions relating to the text and share their ideas before writing their responses individually.

Answers

2 'crept', 'softly'

3 The children were frightened, and they ran away.

4 His eyes were full of tears.

5 And the tree broke at once into blossom. The birds came and sang on it. The little boy stretched out his two arms. He flung them around the Giant's neck. He kissed him.

6 Spring is a time of new beginnings, so the trees will get new leaves and flowers, seeds and plants will start to grow, flowers will bloom, and the birds and animals will come out.

7 Learners' own responses. They should describe something that could have preceded the extract. For example, they could describe where the giant was or what he was doing before he crept downstairs.

Hint

- This advice relates to question 5 of the *Try this* activity and reminds learners about sentence structure.
- Learners should identify and remove the connectives to make sentences shorter. There should be appropriate punctuation at the end of sentences and new sentences always begin with an upper case letter.

Practise

- These tasks are about understanding prepositions. We can use different prepositions with the same verb and change the meaning of sentences, for example we can use *into, up* or *out* with the verb *break* and it has a different meaning each time.
- Learners make up sentences with different meanings using the verb 'break' and given prepositions. Then they complete sentences by choosing the most appropriate preposition to fill the gap.

Answers

1 'To break **up**', 'to break **out**', 'to break **into**', 'to break **down**'. Learners should use these phrases in sentences, for example: *The car broke down on the way to the city.*

2 **a** into   **b** out   **c** up   **d** down   **e** into   **f** down   **g** with

Cartoon strips

> ### Study Guide (pages 61–63)

Learn

- Cartoon strips (or comic strips / comics /comic books) are a type of story that consists of pictures with speech or thought bubbles to tell us what the characters are saying or thinking. Explain to learners that the pictures show us what is happening in each scene. Sometimes, sounds like 'bang' or 'pop' are added around pictures in action scenes (these are not usually inside bubbles).
- Ask a volunteer to read the speech bubbles in the *Learn* box that describe cartoon strips and explain what puns are. Tell learners to look out for the puns in the next activity.

Try this

- Learners read five cartoon strips (A–E) in the activities across pages 62–63. You could read the cartoon strips together as a class first, asking a volunteer to read each speech bubble or to make the loud sounds.
- There are 12 questions relating to the cartoon strips for learners to complete. These include identifying and explaining puns, and using indirect speech. *Note:* Each new picture in a cartoon strip is called a frame.

Answers

1 The word 'average'.

2 It can mean a normal or not very special person, and in maths, it is the word used for the sum of several numbers, divided by their total, so for example, the average of 4, 5 and 6 is 4+5+6 = 15, divided by 3 is 5.

3 The word 'mean' is another maths word for 'average', and it is also a word for being nasty to someone.

4 The word 'odd'.

5 The word 'odd' can mean strange or different, and it is also a maths term (odd numbers and even numbers).

6 **b** – A homophone

7 He fainted or lost consciousness.

8 To become famous

9 The woman asked if her coffee was too strong.

10 It is a pun (also accept the answer that it is a metaphor).

11 The words 'buzz' / bus are being played with, as the two words sound the same.

Let's revise

- Learners should be able to recognise and construct direct and reported speech. Practise taking quotes from stories or newspapers and converting them between direct and reported speech.

- There are many prepositions in the English language. These may only be small words, sometimes just two letters, but they are crucial to the meaning of sentences. We can categorise prepositions into groups including prepositions of location, of direction, of time, and of manner, to name a few.

- Put learners into groups, give each group one category of prepositions and challenge them to recall as many prepositions as they can and to write example sentences using them.

Top tips

- Encourage learners to read a little English every day. This technique will help to widen their vocabulary, deepen their understanding of sentence structure, and consolidate tenses and grammatical concepts.

- Reading a variety of genres including stories, poetry, fiction and non-fiction texts can be an enjoyable and rewarding way to develop langue skills. Learners could write reports or reviews of what they read and share their opinions with others.

Let's go! Worksheet 15: Red and the Wolf

Worksheet 15 provides further reinforcement and practice of reading for meaning. Learners need to go back to page 44 of the Study Guide to read *Red and the Wolf* and then answer questions about it.

Answers

1 She was bored.

2 Not to talk to strangers

3 She called the security company to catch the wolf.

4 So the wolf would not see them.

5 They are so big.

6 He ran quickly towards the door.

7 'Quick as a wink' and 'In one swoop'

8 Tea and cakes

9 A traditional story usually with a moral or message

10 Learners' own responses. They should write no more than one paragraph as a summary, including only the key details rather than retelling a folktale.

Worksheet 15

Red and the Wolf

Reread the story of *Red and the Wolf* on page 44 of the Study Guide. Then answer these questions.

1 Why did Red go to visit her gran?

   _____

   _____

2 What advice did Mum give Red?

   _____

   _____

3 Whose number did Red call and why?

   _____

4 Why did Red and Gran hide in the cupboard?

   _____

5 What compliments does Wolf give about Gran's eyes and ears?

   _____

6 Wolf 'bolted' for the door. What does this mean?

   _____

7 Find two phrases in the text that mean something happened very fast.

   _____

8 What food and drink does everyone enjoy at the end of the story?

   _____

9 What is a folktale?

   _____

10 Write a summary of a folktale that you know.

   _____

   _____

   _____

   _____

   _____

   _____

Section 3: Reading for detail

 Study Guide pages 64–67

 Key skills: Identifying details in advertisements and diary entries.

Advertisements

Study Guide (pages 64–65)

Learn

- Recognising details is a key skill in language learning. When we read for details, we focus on picking out the important information. Advertisements usually catch our attention with pictures or big bold writing, but it is also important to recognise the details.

- Learners may be interested to hear the idiom: 'Always read the small print'. Ask them what they think this means. It means to pay attention to the finest details so that we do not miss anything important.

Try this

- Learners read an advertisement for RollerRide and answer questions about it.

- Put learners into small groups or pairs to read the advertisement together first. Ask them to look at the pictures and to read all the text, in every part of the advertisement.

- Then learners share their ideas to answer the questions before writing their responses individually.

Answers

1 A roller skating park called RollerRide

2 a Learners' own answers. Answers will vary depending on their interpretation, for example: *RollerRide is open. RollerRide is fun. This will be a huge amount of fun.*

   b Yes, they show people enjoying themselves and the roller skates picture shows quality, safe equipment.

   c Big words grab or catch the reader's attention.

   d safe, friendly, dedicated, well-nourished

   e You book online and each session is two hours long.

   f You can roller skate, have a birthday party or function, have a meal and listen to music.

   g You can rent roller skates.

   h Learners' own responses. They should give a 'yes' or 'no' answer about whether they would like to go to RollerRide and they should give an explanation for their choice. For example: *Yes, because it looks fun; No, because I don't like a lot of speeding people all around me.*

   i Learners' own responses. For example: *Grandparents can take grandchildren and it will be cheaper if they don't have to pay; pensioners have less money because they no longer work; pensioners have more time as they no longer work, so they could roller skate more often.*

Diary entries

Study Guide (pages 66–67)

Learn

- Here, learners continue to develop their skills at reading for detail as they read a diary entry written by a reluctant writer! Together with learners, look at the meaning of the word 'derogatory' explained in the glossary box. Ask learners to think of synonyms, for example: 'unkind', 'critical', 'offensive', 'mean', 'hurtful'.

- The writer mentions Jane Austen, and learners can see her portrait on page 67 in the Study Guide. As an extension activity, learners could do some further research at home to find out more about Jane Austen such as where and when she was born and what she wrote.

Try this

- Learners read the diary text first and then answer questions. Tasks include identifying factual information from the text and identifying reasons for things described in the text.
- Reading could be done individually, in pairs or groups, or as a class.
- Some vocabulary is explained in the glossary box, remind learners to use a dictionary to look up the meaning of any other words they are not sure about.

Answers

1 B

2 B

3 Learners' own responses, which should express the emotions on the faces, for example: 'Ugh! Oh no!' or 'Oh no! Why? I can't believe it!' (Or any such combination)

4 The word 'nuts' can be offensive to some people. If you say someone is nuts, it implies that they are not of sound mind, which is not a nice thing to say, as mental health issues are serious and important.

5 In the 2020 global Covid pandemic lockdown: 'So I can 'journal' to 'express myself **during lockdown**'.'

6 B

7 Play on their phones while putting a picture of themselves on their computers, so their teacher could not see that they were not listening to him or her.

8 'I. Am. So. Bored.'

9 The cat is stupid because it likes to sleep in the washing machine and so runs the risk of getting washed.

10 Fresh air and exercise

Let's revise

- Use the RollerRide advertisement to look more closely at verb forms. What different types of verbs are used? Can learners identify imperative verbs? Infinitives? Gerunds?
- Encourage learners to talk about why different verb forms are used in different places.

Top tips

- This kind of diary entry is known as 'journalling'. Journalling is where a person writes not only about the things they do in a day, but also how they felt or what they thought.
- More confident learners could try some journalling over a period of a few days. Then they could bring their journals to class to share and compare with others.

Let's go! Worksheet 16: Advertisements

Worksheet 16 provides further reinforcement of understanding advertisements.

Answers

1 A cream for spots / pimples

2 To engage the reader, to make the reader stop and think.

3 People that suffer from spots / teenagers

4 There will be half as many spots in one week.

5 Twice a day

6 At the pharmacy / chemist

7 People know what the product looks like / they will recognise it in the shops / it appeals to young people like the girl on the packaging / it is brightly coloured

8 Learners' own responses. Adding a price to the advert or making text bigger could be a suggested improvement. Learners may say it is good because it is engaging and enthusiastic to make people want to buy it.

Worksheet 16

Advertisements

> Fed up with spot breakouts? Don't want to leave
> your bedroom because of your pimples?
> The solution is here: **Spot Out** cream!
> Guaranteed to halve your number of spots
> after just one week.
> Apply directly to spots
> morning and night for best results.
> Find it at your local pharmacy.
> No more hiding away. Buy **Spot Out** today!

1 Skim read the advertisement. What is being advertised?

_____

2 Why do you think the advertisement asks questions?

_____

3 Who might buy this product?

_____

4 What promise does the advertisement make?

_____

5 How often can a person use this product?

_____

6 Where is this product for sale?

_____

7 How does the picture help advertise the product?

_____

_____

_____

8 Do you think this is a good advertisement? Could it be improved?

_____

_____

_____

_____

Section 4: Reading instructions

 Study Guide pages 68–69

 Key skills: Reading and following instructions; answering questions about instructions.

Instructions

> **Study Guide** (pages 68–69)

Do you remember?

- Learners hear instructions every day in many different places and situations. Ask volunteers to give you an example of an instruction that they have heard today. Remind learners what instructions are, using the bullet points on page 68.
- Talk to the class about adverbs that are used for sequencing, in other words, to put events in order.
- Talk to the class about the command forms of verbs – these are called imperatives. You could ask volunteers to find examples of imperatives in the First Aid instructions in the *Try this* activity.

Try this

- Learners may know, or may have performed, First Aid. Ask learners to talk about the skills they have or the experiences they have had.
- Put learners into small groups to read the First Aid instructions together. They should take turns to read the text aloud. The tip box gives a useful mnemonic for First Aiders to use to remember what to do in an emergency situation.
- After reading, learners answer questions about the First Aid advice. The final question asks for an opinion and a reason. Make sure that learners expand on their opinions, giving details and reasons.

Answers

2 a Check for hazards and make sure that you are safe.

  b Step 1 is important because you can't help someone else if you are not safe or if your own life is in danger.

  c Hazards, Hello, Help, Airway, Breathing, Circulation

  d External hazards, such as fire, petrol on the road, animals (dogs); biological hazards, such as blood or airborne viruses such as Covid-19. (Any three)

  e You need to say hello into each ear and gently tap each shoulder to check if someone is conscious.

  f No, you shouldn't leave the patient. If you have a cell phone, then you should phone for help. Otherwise you should ask someone to call emergency services.

  g A, B, C – check the airway, the breathing and the circulation.

  h Learners' own responses.

Practise

- These questions are about understanding the meanings of words relating to First Aid. First, learners match words to meanings, and then they write sentences using each word. To provide differentiated support for less-confident learners, ask them to write fewer sentences while more confident learners should write a sentence for all of the words.
- This activity could be extended by revising other related vocabulary such as parts of the body, health and well-being, or vocabulary to do with hospitals and the emergency services.

Answers

1 a Emergency – A sudden state of danger needing immediate action

  b Initial – First

  c Procedure – A way of proceeding or performing a task

d Hazard – A danger or risk

e Conscious – Awake or aware

f Circulation – Movement of the blood to and from the heart

2 Learners' own responses, using each word from question 1 in a sentence.

Let's revise

- It can be helpful to revise adverbs and instructions by talking about something we know well, enjoy or do regularly, for example, explaining a skill or hobby that we have. You could ask learners to explain to a partner or in small groups how to play a sport or musical instrument, or how to do something else they enjoy.

- Remind learners that they should use imperative verb forms for instructions and sequencing adverbs to explain the steps in order.

Top tips

- To engage learners with the text, they could try to do some basic First Aid in the classroom. For example, they could practise putting each other in the recovery position or practise some types of bandages such as making slings.

- When practising, learners will need to use imperative verb forms to tell their partner what they need to do.

Let's go! Worksheet 17: Instructions

- Use Worksheet 17 for consolidation of recognising and understanding instructions, in this case in a recipe text.

- Learners read a recipe and identify parts of speech including imperative verb forms, adjectives and adverbs.

- In the answers below, adjectives that end in '-ing' (denoting their purpose) are accompanied by the noun they describe for clarification.

Answers

1 Sift, add, mix, leave, heat, stay, pour, use, scoop, put, cook, use, add, enjoy.

2 baking (powder), white, mixing (bowl), frying (pan), little, hot, big, small, golden-brown.

3 Thoroughly, slowly, gently, carefully.

Worksheet 17

Instructions

Easy pancakes

For this recipe you will need:

$\frac{1}{2}$ cup flour

3 teaspoons of baking powder

a pinch of salt

$\frac{1}{2}$ tablespoon white sugar

$1\frac{1}{4}$ cup milk

1 egg

3 tablespoons of oil (plus extra for frying)

What to do:

1 Sift the flour, baking powder, salt and sugar into a mixing bowl.

2 Add the milk, egg and oil to the mixture and mix thoroughly.

3 Leave to stand for a few minutes.

4 Heat a frying pan. Stay safe – ask an adult to help you light and use the stove!

5 Pour a little oil into the pan. Stay safe – if the pan is too hot, the oil may spit!

6 Use a big spoon or a small cup to scoop mixture from the bowl. Put each spoonful or cupful slowly into the pan.

7 Cook gently on both sides until the pancake is a golden-brown colour.

8 Use a spatula to remove the pancake carefully from the frying pan.

9 Add a topping of your choice.

10 Enjoy!

1 Circle or highlight all of the command verbs in the recipe.

2 List all the adjectives in the recipe.

_____

_____

3 List all the adverbs in the recipe.

_____

_____

Section 5: Reading an argument

Study Guide pages 70–71

Key skills: Skimming; reading an argument; recognising points of view.

Arguments

Study Guide (pages 70–71)

Do you remember?

- In Unit 1, Section 6, learners explored arguing different viewpoints. Here, they will explore arguments further by reading an argument and recognising the different viewpoints that are expressed.

- Before starting the activities, have a class discussion about arguments, covering the information in the bullet points. Arguments are not always about raised voices and shouting to make your point; an argument is a type of conversation, and it is important to respect and listen to the opinions of other people, regardless of how different they may be from your own views.

Try this

- In this activity, learners read a transcript. A transcript is a word-for-word written record of a conversation, in this case a radio interview. Explain that Fred Medley is the radio host, and Grace and Joe Cluck are the two people being interviewed.

- First, ask the class to skim read the transcript. Skim reading means to quickly read over the text to understand the general idea of what the text is about. Then ask volunteers to share their ideas for the first question. Next, allow learners to work in groups of three to read the text together; each learner could read the part of a different person.

- Have dictionaries on hand for learners to look up any tricky vocabulary. Remember, transcripts are word-for-word accounts, so some of this language will be unfamiliar to learners. Reassure learners that they do not need to understand every word to follow the argument or answer the questions.

- After reading the transcript, learners can talk through the questions together before writing their individual responses. Encourage learners to talk about their own opinions regarding vegetarianism. Questions in this activity include comprehension, giving quotes, recognising connectives and interpreting metaphors, before a final task to write a short paragraph giving an opinion. *Note:* A metaphor is when something is used to represent something else, for emphasis or to make it more visual.

Answers

1   Fred Medley, Grace, Joe Cluck. They are arguing about being a vegetarian or a meat eater.
2   Vegetarianism
3   A
4   A
5   He needs to moderate the discussion and make sure each person gets a turn to present their views.
6   Learners' own responses. Accept any suitable sentence said by Grace, for example: 'I would ban killing animals, that's for sure.'
7   Learners' own responses. Accept any suitable sentence said by Joe, for example: '… the fact is that humans are meat-eaters, and we have been for centuries.'
8   Learners' own responses. Accept all correct explanations, for example: It means when it's time to slaughter the animals for their meat.
9   Learners' own responses. Accept all correct explanations, for example: Don't force us to accept your opinion or message.
10 'and', 'because'
11 Learners' own responses.

Let's revise

- Revise vocabulary for vegetables by completing a fruit and vegetable A–Z. This could be done verbally as a ball game, or by writing a list either individually or on the board as a class. For how many letters of the alphabet can learners name a fruit or vegetable that begins with that letter? How many fruit or vegetables can learners name for each letter?
- Alternatively, combine revising this vocabulary with revising colours by asking learners to categorise fruits and vegetables according to colour.

Top tips

Some English metaphors are common in everyday speech and language, and many idiomatic phrases are based on metaphors. You could explore some common metaphors and idioms in more detail with the class. For example, tell the class phrases such as 'a blanket of snow' and ask them what they think such metaphors mean.

Let's go! Worksheet 18: Different points of view

- Use Worksheet 18 for further reinforcement and practice about understanding arguments.
- Learners read four statements and make notes of arguments for and against each one. Then they share, compare and discuss their arguments with a partner or in a small group.
- This task could be extended by selecting one or more statements and using them for a class debate.

Worksheet 18

Different points of view

Here are some statements that people often have different opinions about. Do some research if you need to and make notes of arguments for and against each statement. Then share and compare your arguments with a partner or in a small group.

1 People should give up meat and become vegetarian.

Arguments for	Arguments against

2 People should stop driving cars and ride bicycles instead.

Arguments for	Arguments against

3 People should reuse rainwater in their homes.

Arguments for	Arguments against

4 People should only use renewable energy for electricity.

Arguments for	Arguments against

Section 6: Reading poems

 Study Guide pages 72–75

 Key skills: Reading and understanding a range of styles of poetry.

Poems

> ### Study Guide (pages 72–75)
>
> ### Do you remember?
>
> - Poetry takes many forms and often uses comparisons to have a specific effect on a reader. Read the explanations and examples of the two types of comparisons often found in poems to the class or ask for volunteers to read the examples.
> - Also note the features of poems. Poems are written in verses, lines are short and sentences can be split from one line to the next instead of continuing in a long line.
>
> ### Practise
>
> - Learners read the poem *The shepherd dreams*, by Fiona MacGregor. Put learners into pairs or small groups to read the poem together. They should each take turns to read parts of the poem, perhaps a verse at a time. You may prefer to read this to the class first, so that learners can hear your intonation and listen to how you use the punctuation as a guide for when to breathe and when to pause.
> - After reading, learners respond to questions about the poem to assess their understanding. The questions can be discussed in pairs, groups or as a class prior to learners answering them individually.
>
> ### Answers
>
> 1 He is dreaming about life in the city.
> 2 The poet thinks the warm sun and the beautiful green hills are paradise: 'And paradise lies at his feet.'
> 3 The sounds the sheep make are like people mumbling.
> 4 It is sudden:' Sun slices above the hills.'
> 5 Learners' own responses, for example: Shepherd – a sandwich, porridge, any other reasonable answer; lamb – milk.
> 6 He wants cars, phones, lights, noise and busy streets.
> 7 Learners' own responses. Answers will vary depending on their interpretation.
> 8 Yes: Knees and breeze (lines 3 and 6), care and air (lines 9 and 12), streets and feet (lines 15 and 18).
>
> ### Practise
>
> - Have learners read the poem *The blind man waits* by Fiona Macgregor either in pairs, groups or as a class. Again, you may prefer to read the poem aloud yourself first as an example for learners to pick up on reading techniques.
> - Some of the vocabulary for this poem is explained in the glossary. Allow time for learners to ask about or look up any other vocabulary they are unsure of.
> - Before moving on to the questions, talk about the picture with the class. What can learners see? How would they describe the scene?
>
> ### Answers
>
> 1 He is waiting for some form of human connection – all quotes from the last verse are acceptable, for example: 'Waiting for a word' / 'A moment of connection'
> 2 Learners' own responses, depending on their interpretation. They may say he feels lonely because the last two lines say he wants connection with others.
> 3 beep beep, whoosh, rumble (any two)

4 The poet describes him as an island in the street, so he is separated from other people.

5 It sounds like small pieces of metal hitting the roof.

6 Learners' own responses, for example: Probably in some ways, as your hearing and sense of smell will be better.

Try this

- Ask learners to read the poem *The hot wind* by Fiona Macgregor. This short poem describes things we see and feel when a hot wind blows, with leaves blowing about and laundry hanging on washing lines to dry.

- It is a short poem, with lots of imagery for learners to discuss in pairs or groups. Learners can then share their ideas for the questions before writing their responses individually.

Answers

1 rattle, run, road

2 The leaves are being blown along the road in the wind.

3 Yes, because that is the sound that wet sheets make in the wind.

4 There's electricity in the air, like there often is with a hot wind, and the static makes her shirt stick to her skin.

5 It emphasises that the wind keeps blowing; the relentlessness of the wind.

6 Learners' own responses.

Let's revise

- Use the poems in this section to do some phonics practice. Look at the vowel combinations and consonant combinations in the *The hot wind* poem, for example.

- Ask learners to repeat lines and focus on the sounds. Ask them to identify lines that have words with long or short vowel sounds or silent letters.

Top tips

- Question 6 in the *Practise* activity on page 75 requires learners to use their critical thinking skills to put themselves in someone else's position. You could encourage a wider discussion about disabilities and ask learners to share their opinions on matters such as disabled access in school and in the local community, for example.

- As with *The hot wind* poem on page 75, learners could take inspiration from the weather and write their own short poem. They could use the type of weather as the basis for an acrostic poem. For example, each line could begin with the sequential letters of the word 'sunny' or 'thunder'.

Let's go! Worksheet 19: Poems

- Worksheet 19 provides further reinforcement and practice reading poetry.

- Learners read a poem that has lots of imagery. Then they answer questions about the poem and draw their own picture of the scene they have in their imagination after reading it. Learners will need colouring pencils.

Answers

1 Spring

2 Summer

3 Crickets, they are chirping

4 Warm and cool, hot and cold

5 Fresh, morning (dew), wet, ginger (biscuits), sweet, green, bright, long. Do not accept adjectives given in question 4. Do not accept 'gentle' – this is an adverb describing the chirping of the crickets.

6 Yes, the adjectives are all happy, for example: 'bright', 'warm'

7 Pictures should contain elements from the poem such as sunshine, grassy fields, crickets, or milk and biscuits.

Worksheet 19

Poems

> **Springtime**
>
> Springtime smells like fresh morning dew
> And feels like cool wet grass between my toes.
> Springtime is the warm sun against my face
> And the gentle chirp of the crickets in the fields.
> Springtime is Granny's ginger biscuits,
> Hot and sweet with cold milk to dip.
> Springtime is always green and bright;
> It promises summer is almost here
> And I dream of those long hot days to come.
>
> *Jennifer Peek*

1 Read the poem. What season is the setting for the poem?

   _____

2 What other season is mentioned?

   _____

3 Which insects are mentioned and what are they doing?

   _____

4 Find two pairs of adjectives in the poem that are antonyms (opposites).

   _____

5 List at least four other adjectives from the poem.

   _____

6 Do you think the poet likes springtime? Quote words from the poem to support your answer.

   _____

7 In the space below, use colouring pencils to draw a picture of the scene you see in your imagination after reading the poem.

Section 7: Reading idiomatic expressions

 Study Guide pages 76–77

 Key skills: Reading and understanding a range of English idioms.

Idioms

Study Guide (pages 76–77)

Learn

- In Section 7, learners explore idiomatic expressions and their meanings. Importantly, they deepen their understanding of the difference between idiomatic expressions and facts, as explained in the *Learn* box.
- Ask a volunteer to read the speech bubble about just how common idioms are in English. You could begin by asking learners to share any idioms they can recall with the class.

Try this

- First, there are eight English idioms that learners need to match to their meanings. Encourage learners to try to visualise the idioms and to look for clues in the phrases to help them to work out the meanings.
- Next, learners choose three idioms and use them in sentences. To provide differentiated support, ask less-confident learners to write fewer sentences, even if it is just one.

Answers

1 a To live within one's income

c Something that happens very rarely

e In the public eye, getting a lot of attention

g Start at the beginning with no extra help

b To begin again, and plan to do it better

d Be the first one to begin

f In the same set of circumstances

h Avoid coming into contact with a person

Practise

- In this *Practise* activity, learners read a dialogue. Like a transcript, a dialogue is also a written record of a conversation. However, a transcript is a record of a conversation that has happened, been listened to and then written down, while a dialogue may not be a real conversation; it could be a script that has been written before it is read by actors.
- The dialogue in question 1 contains ten idiomatic phrases for learners to find. First, put learners into pairs to read the dialogue together, each learner reading one of the parts. Tell them to underline the idioms that they find. Can they find all ten phrases?
- In question 2, learners are challenged to rewrite the dialogue and remove the idioms. It will be useful for learners to have some rough paper to write a draft dialogue first. Then they can swap drafts with a partner to check each other's work and provide feedback before they write up their final dialogues neatly.
- Question 3 is based on a short reading passage about future goals; Rosa explains why she wants to be a fashion designer.
- Learners need to identify and explain the idioms that Rosa uses. When reading, learners will notice that there is no punctuation at all in the text. The final task is for learners to add the punctuation to the paragraph, which includes signposting the start and end of sentences using punctuation, as well as punctuation within sentences.

Answers

1 Ted: Well, <u>it's just not cricket</u>, is it? I mean, I may <u>be a bit long in the tooth</u> but I know a <u>two-faced</u> person when I see one!

Jack: You <u>got out of bed on the wrong side</u> this morning! Sounds like you've <u>got a bone to pick</u> with someone.

Ted: That Gloria, she's <u>full of airs and graces</u>, but she's just like her mother underneath it. <u>All hot air!</u> <u>When the chips are down</u> you'll see – she'll <u>show her true colours</u>.

Jack: Blimey, you really <u>have a bee in your bonnet</u> about her, don't you?

2 Learners' own response, for example:

Ted: Well, it's just not fair, is it? I mean, I may be old, but I know when someone is not saying what they mean!

Jack: You are grumpy! It sounds like you have a problem with someone.

Ted: That Gloria, she thinks she is better than everyone, but she's just like her mother – lots of talk but no real substance. When things get difficult, you'll see what she is really like.

Jack: Gosh, she really irritates you, doesn't she?

3 a To do something again from the beginning

b These are the things that are necessary to make something happen, but we don't see them.

c Behaving in a particular way, usually different to the way you usually behave

d I'm going to be a fashion designer with a difference. My plan is not to start from scratch, but to use old clothes as my starting point. I'll cut them up and reuse the good pieces of fabric to make stunning new garments. I think I'll call it Behind The Scenes, or True Colours. What do you think?

Let's revise

As a combination of reading and writing skills, learners could write a paragraph about their own future goals. They should demonstrate their writing techniques including planning, drafting, finalising and proofreading, and they could try to include idiomatic phrases as well.

Top tips

Learners could practise speaking skills by performing the dialogue from page 77 with a partner for the class or in front of others in a group.

Let's go! Worksheet 20: Idioms

- Worksheet 20 provides further practice with idioms.
- First, put learners into pairs or groups to discuss the meanings of common idioms. Then, they identify idioms in sentences and finally they explain what an idiomatic expression is by giving an example.

Answers

1 Expressions on the left matched with their definitions on the right:
- to feel like a million dollars – to be very, very happy about something
- to be as quiet as a mouse – to make so little noise that people
- to be as brave as a lion – to face a difficult situation with courage
- to know something like the back of your hand – to be familiar with every detail of something
- to be as free as a bird – to be able to do anything you please whenever you want to
- to sing like an angel – to have a beautiful voice
- to be as strong as an ox – to be able to lift things that are very heavy

2 a Lisa is a great singer, but she doesn't like to be in the limelight.

b I can't fix it, so I'll have to start from scratch.

c Let's break the ice by playing a game together.

d Honestly, you and your brother are two peas in a pod!

e We are all in the same boat, you know. There's no point complaining!

3 Learners' own responses. Learners should give an example of a sentence with an idiomatic expression and one with a fact to illustrate the difference, for example: 'She broke her finger.' / 'She broke her heart.'

Worksheet 20

Idioms

1   Work with a partner. Read and discuss the English idioms below. Draw lines to match each idiom on the left with its correct meaning on the right.

to feel like a million dollars	to have a beautiful voice
to be as quiet as a mouse	to be able to do anything you please whenever you want to
to be as brave as a lion	to be able to lift things that are very heavy
to know something like the back of your hand	to make so little noise that people hardly know you're there
to be as free as a bird	to be very, very happy about something
to sing like an angel	to face a difficult situation with courage
to be as strong as an ox	to be familiar with every detail of something

2   Underline the idiom in each sentence.

   a   Lisa is a great singer, but she doesn't like to be in the limelight.

   b   I can't fix it, so I'll have to start from scratch.

   c   Let's break the ice by playing a game together.

   d   Honestly, you and your brother are two peas in a pod!

   e   We are all in the same boat, you know. There's no point complaining!

3   Write a sentence to explain the difference between idiomatic expressions and facts. Give examples to support your answer.

   _____

   _____

   _____

   _____

   _____

   _____

Check your understanding

Study Guide pages 78–79

- These end-of-unit exercises can be used to check how well learners understand aspects of reading that have been covered throughout Unit 3. There are two parts, based around two texts that are taken from the same story, *Year 6 at Greenwicks* by Adam and Charlotte Guillain.
- The questions include comprehension to assess learners' understanding of the text , as well as taking quotes from the texts, identifying idiomatic expressions, using reported speech and expressing opinions based on the texts.

Answers

1 a 'hurried'

b They have been friends since the first day of pre-school.

c It means that a person has an angry or unhappy expression on their face.

d The trousers are too short, because he has grown.

e C – white socks and black shoes

f Carter's mum and dad are separating, and his dad is moving out of their home.

2 a It means that they are signing their names in the attendance register.

b They are in Year 6 and need to set an example for the younger learners.

c Mrs Wilde asked the boys to try and get to school on time tomorrow.

d There is a new teacher, Mr Ali, and he is putting learners into different reading groups from those they have been in previously.

e They protest because they have been together in the same reading group since Year 1.

Unit 4 · Use of English

Section 1: Grammar

Study Guide pages 80–93

Key skills: Asking and answering questions and making suggestions; using verb forms in a range of past and present tenses; talking about the future in different ways; beginning to understand conditionals; recognising active and passive sentences; using direct and reported speech.

Asking questions

Study Guide (pages 80–81)

Learn

- Grammar is the nuts and bolts of languages. Grammatical rules and patterns hold words together so that they make sense and can be understood by others. Section 1 of Unit 4 revises some key grammatical concepts, starting with asking questions.

- Learners know that questions can be asked in many different ways in English. Some questions use standard question words, some questions are tagged onto the end of statements, some questions use auxiliary verbs, sometimes we change the order of words to create a question, and sometimes we can even change a statement to a question just by using a different intonation.

Try this

- This activity is about recognising how questions are constructed. Learners look at eight questions and identify how they have been formed from a choice of options.

- Once learners have completed the activity, go through the answers with the class, identifying the question word, verb (specifically the auxiliary verbs 'can', 'have', 'do') or the word order in each case.

Answers

1 Question word (how)   2 Word order

3 Question word (why)   4 Question word (where)

5 Verb (can)   6 Question word (who)

7 Verb (have)   8 Verb (do)

Practise

- This exercise encourages learners to construct questions of their own. The first word is provided for learners to complete each question. They are instructed to try to construct at least four questions; you could challenge more confident learners to write all of the questions.

- Extend this activity by putting learners into pairs or small groups to share, compare and discuss their questions. How are they similar and how are they different?

Answers

Learners' own responses, for example:

- What equipment do you need?
- How much does it cost?
- Where can you do white-water rafting?
- Is it fun?
- Why is it called 'white'-water rafting?
- Do I need to be able to swim?

Try this (page 81)

This activity checks learners' understanding of questions as they match questions to possible answers. Learners should look for clues to help them, for example a question that asks 'where' needs to have a location in the answer, a question that asks 'why' needs to have a reason in the answer. The *Hint* box tells learners to look out for key words to help to do this.

Answers

- How can I learn to spell this word? You have to practise writing it and looking at it!
- Why do we have to write tests? Because they help us understand what we have learned.
- Do you know how to play chess? Yes, I do. Should I show you how?
- Where is the nearest bus stop? It's down the road, on the right.
- Where were you born? I was born in Kuala Lumpur.
- Are you busy right now? No, I am not.

Making suggestions

> **Study Guide (pages 81–82)**

Learn

- Just as there are many ways to ask questions, so there are also a variety of ways to make suggestions. Making a suggestion is, in fact, another form of asking a question.
- Ask volunteers to read the speech bubbles in the *Learn* box and then go through the different ways to form suggestions in the bullet points, using the suggestions in the speech bubbles as examples.
- Learners may need reminding that the infinitive is the part of the verb with 'to' in front.

Practise

- This task revises and consolidates making suggestions. Learners choose the correct words from two options to complete each suggestion.
- Encourage learners to think about whether they need an infinitive, a gerund or a noun to make the suggestion make sense.

Answers

1 going

2 having

3 to listen

4 going

Try this

This activity gives learners the opportunity to construct sentences to create a dialogue. Learners could work in pairs to come up with the suggestions together. Then they can perform their version of the dialogue for the class or in small groups.

Suggested answer

Maya: It's Ali's birthday on Saturday. Let's **make** him a surprise.

Paul: That's a good idea. What about making **a birthday cake**?

Maya: Yes. Or how about **buying a gift**?

Paul: How can we do this? Would you like to **buy the ingredients for the cake**?

Maya: Fine. And how about you? What will you **wrap the gift**?

Paul: I'll **wrap it and bring it to the party**.

Maya: Great. Let's **go**!

Paul: **Sure.**

Present simple tense: active and passive

Study Guide (pages 82–83)

Do you remember?

- Ask learners to read this information for a reminder about how we use the present simple tense in English. Ask a volunteer to read the speech bubbles and then allow time for learners to ask any questions that they may have.

- Note that the text points out that the present simple tense is not used to describe what we are doing right now. To do that, we use the present continuous tense and learners will revise that a little later in Section 1.

Practise

- Begin by asking learners to look at the picture and to tell you what they can see. Do they recall the word for the tall metal structures? (wind turbines) Ask: *What do they do? What are they for? What is the weather like in the photo?*

- Then tell learners to read the paragraph and circle the verbs in the present tense.

- You could extend this exercise by drilling vocabulary related to natural resources. Learners could also make a mind map of related vocabulary in their notebooks.

Answers

use, create, are, turns, makes, spin, gives, produces

Do you remember?

- Ask learners to read this information to revise the difference between active sentences and passive sentences in English, and how to form passive sentences.

- Make sure that learners demonstrate their understanding of past participles by asking them to give some examples (regular and/or irregular) before moving on to the *Try this* activity.

Try this

- This activity reinforces learners' understanding of active and passive sentences. Learners read four active sentences and convert them into passive sentences.

- Ask learners to pay attention to the *Hint* box, which reminds them about irregular past participles.

Answers

1 The vegetable seeds are planted by Jo in the spring.
2 The arms of a wind turbine are turned by the wind.
3 Smartphones are used by many people.
4 Every Saturday, cakes are made by Shameela / Cakes are made by Shameela every Saturday.

Past simple tense: active and passive

Study Guide (pages 83–84)

Do you remember?

- This information reminds learners about the past simple tense, which we use to describe completed events in the past. Like the present tense, sentences using the past simple tense can also be active or passive, as the examples show.

- Ask volunteers to read the examples. The insect in the photograph is a mosquito, which has a tricky spelling – spell and write it letter by letter on the board, then rub it off and ask learners to say or write it and check their spelling. This could be developed further into more drilling for insect vocabulary, or for other words that have the 'qu' combination. For example, learners could name and describe different types of insects and bugs that they find in the school grounds.

Practise

This task reinforces how to construct sentences in the past simple tense. Learners choose the correct words from three options to complete the sentences. The options are all verbs, so encourage learners to think about who or what is doing the action in each case to help them to make their choice. Then they underline any passive sentences – there is only one (question 3).

Answers

1 was	**2** stayed	**3** was written (passive sentence)
4 downloaded	**5** brought	**6** started

Try this

- In this activity, learners can demonstrate not only their understanding of the past simple tense, but also the range of verbs that they know and can use in sentences. Learners read a paragraph about someone who moved to a new house and complete the gaps with appropriate verbs.

- There are no fixed answers for this activity, because the key point is for learners to use verbs that make sense in this context. Some gaps could be filled by several verbs, for example, the first gap could be 'went' or 'moved' or 'travelled' or 'relocated'. Other gaps really only have one possible option that makes sense, for example: 'We hardly **had** any room to sit!' After completing the activity, learners can swap answers with a partner to share and compare their choices.

Answers

Learners' own responses, for example:
Last week, we **moved** to a new house. It was quite exhausting, but in a way it **was** fun too! First, all our things were **stored** in boxes. Then the movers **came** to collect the furniture and the boxes. We **put** our suitcases in the car, with other small things. We hardly **had** any room to sit! Then we **travelled** to the new house. What a day!

The present continuous tense

> **Study Guide** (pages 84–85)

Do you remember?

- As mentioned earlier in Section 1, the present simple tense is not used to describe what we are doing right now. To do this, we use the present continuous tense, as this information explains.

- Ask a volunteer to read the speech bubbles.

- Ask volunteers to give example sentences using the present continuous tense to describe what they are doing right now. For example: 'I am listening to my teacher.', 'I am learning English.', 'I am reading this book.', 'We are sitting in our classroom.', 'We are breathing.', 'We are smiling.'

Practise

- This activity consolidates understanding of the structure of the present continuous tense. Learners choose the correct words from three options to complete the sentences.

- Ask learners to focus on the components of the present continuous tense. They should try to identify the present tense of 'be' and the present participle that go together in each sentence.

Answers

1 are	**2** sending
3 going	**4** is ploughing
5 is taking	

Try this

Learners complete the gaps in sentences to make a dialogue between friends. They could do this individually and then share and compare with a partner or they could work together to construct the dialogue and then perform it for the class or in small groups.

Answer

Learners' own responses, for example:

Mara: Hi Shabir. You look happy! What's going on?

Shabir: **I am** excited. We're **going to visit** a game park this weekend.

Mara: Oh, wow! **That sounds** wonderful! You **are** so lucky! When are you_**going**?

Shabir: Tomorrow evening. I'm **taking** my new phone of course.

The past continuous tense

> ### Study Guide (pages 85–86)

Do you remember?

- Just as we use continuous verb forms in the present, we also use them in the past. The past continuous tense describes something that was happening in the past, as this information explains.
- Ask volunteers to read the example sentences aloud and to identify the past tense of 'be' and the present participle in each case.

Practise

- In the first part of this activity, learners match beginnings and endings of sentences. Then, in the second part of the activity, they write the complete sentences neatly.
- Tell learners that they should look at the pictures for clues to help them to decide which parts of the sentences match up.

Answers

1 a I was reading peacefully before you arrived.

  b They were driving around in the game park when they saw a herd of elephants.

  c Peter was making supper when he heard the news.

  d Debbie was filming us as we were dancing.

Try this

- This is a gap-fill activity. Learners use the verbs provided in brackets to fill in the gaps by converting the verbs to the past continuous tense. You may need to remind learners to also include the past tense of 'be', which is not included in the brackets.
- The *Hint* box advises learners to check their spelling, because sometimes verbs need to remove an '-e' or double a consonant in the continuous form.

Answers

I was lying on the riverbank. The sun **was shining** and it was quite hot. I **was thinking** about getting into the cool water, when I saw the hippo. First, I just saw two enormous eyes. Then I saw the nose. As I **was collecting** my things to move away, another hippo appeared. Soon I **was running** and the hippos **were getting** closer! As I reached the car, they they **were bellowing** loudly. I was happy to be near the car!

Present perfect tense

> ### Study Guide (pages 86–87)

Do you remember?

- This tense is used mostly for talking about experiences and is often combined with adverbs to give extra details.
- Ask a volunteer to read the speech bubbles to show how to ask and answer a question in the present perfect tense.

Practise

- This activity strengthens understanding of the structure of the present perfect tense. Learners choose the correct words from three options to complete sentences.
- Tell learners to focus on the components of the present perfect tense. They should try to identify if the sentence needs *have* or *has* and to think about the past participle that is needed in each case.

Answers

1 has lived

2 have seen

3 has never been

4 has been

Try this

- This is a game for learners to play with a partner or in groups. Each pair or group will need a timer, or you could set a timer for all groups to follow.
- Learners ask and answer questions in a fixed amount of time. The questions and answers must use the present perfect tense, so it is important for them to think about accuracy and not try to rush, because this can cause mistakes.
- The *Hint* box reminds learners about using *have* or *has* correctly.

Talking about the future

Study Guide (pages 87–88)

Do you remember?

- Before learners read the information in the table, remind the class that there are a variety of ways in which we can talk about the future in English.
- Encourage learners to share what they can recall about how we talk about the future, and ask volunteers to give example sentences that refer to something that will happen in future.
- Read through the information in the table about the different ways in which we can talk about the future and ask a volunteer to read the example in each case.

Hint

- Learners may already have mentioned this in the earlier discussion; suggestions are things that have not happened yet – so it is true to say that suggestions are another way in which we can talk about the future.
- One particular suggestion word that is often used to talk about the future in English is 'shall'. Ask volunteers to give you some more examples of suggestions using 'shall'.

Practise

- This task consolidates learners' understanding of talking about the future in different ways. Learners choose the correct words from three options to complete sentences.
- Tell learners to try to identify the key parts of the sentence structure. Explain that they can refer back to the table in the *Do you remember?* box in the Study Guide for clarification if they need to check how different ways of talking about the future are constructed.

Answers

1 starts

2 will do

3 are going to do

4 won't help

Try this

- This is an open-ended writing activity; the only requirement is that sentences must refer to the future. Learners can use any structure for talking about the future; that they choose, but they should make sure their sentences make sense. Sentences do not have to be about the distant future; they can be about things that will happen later today, tomorrow or this week.

- Learners may benefit from using rough paper or a notebook to plan their sentences first, so they have space to try out sentences in different ways and to edit and change them before writing their final sentences neatly in the Study Guide.
- Extend this activity by putting learners into pairs or small groups to share and compare their sentences, and to talk to each other about their future plans, arrangements and promises.

The first conditional (if clauses)

Study Guide (pages 88–89)

Learn

- Conditionals talk about things that are possible. The first conditional uses 'if' to talk about things that could happen in the present or the future. In later learning, learners will go on to explore other conditional forms that have different usage in English.
- The key point that will help learners to recognise first conditionals at this stage is for them to identify the condition and the result. Ask volunteers to read the example sentences and make sure learners understand the conditions and results in each case.
- What can be a point of confusion for learners here is that, in first conditional sentences, we can swap the order of the clauses and the sentences will still make sense. Ask volunteers to read the new order of the example sentences given in Pia's speech bubble. So, in first conditional sentences, the word 'if' can go either at the start of a sentence or in the middle of a sentence without changing the meaning.

Practise

- Learners can revise changing the word order of first conditional sentences by completing this activity. They read three sentences, all of which begin with 'if'. Then they rewrite the sentences by swapping the clauses, making the result the first clause and the condition the second.
- Remind learners that they must move 'if' when they change the word order of the sentence. It moves to the middle of the sentence when the result is given before the condition, and the comma that was in between the clauses is removed.

Answers

1 You will get fitter if you cycle to school.
2 If you help me with my maths homework, I will teach you how to knit.
3 He won't be so afraid of the sea if he learns to swim.

Try this

Here, learners match clauses together to form first conditional sentences. Then they write the sentences neatly. Encourage learners to look for clues that link the parts of the sentences together.

Answers

1–2 a We will be able to hear the song if you all stop talking so loudly!
   b If you save your money, you will be able to buy something you need.
   c You will see my house if you walk around the corner.

Prepositions that follow verbs

Study Guide (pages 89–90)

Do you remember?

In English, some verbs are always followed by a particular preposition. We call these prepositional phrases. There are no fixed rules or patterns for prepositional phrases, we must learn them on a case-by-case basis.

Practise

- The table in question 2 of this activity lists some of the most common prepositions and gives examples of verbs that use them to create prepositional phrases. Learners should write at least one more example of a prepositional phrase in each case.
- Remind learners that a prepositional phrase is composed of a verb followed by a preposition, and the verb is always in the infinitive form ('to' does not have to be included in these examples).
- The *Hint* box advises learners to refer back to this list when they are doing writing tasks. There are many options for answers. If learners are in doubt about the validity of a response, they should check it in a dictionary.

Try this

This activity deepens understanding of prepositional phrases as learners fill the gaps in sentences with appropriate prepositions. Remind learners that they can look at the table in the *Practise* activity on page 89 in the Study Guide for help, as the *Hint* box advises.

Answers

1 about	**2** with / from
3 with / from	**4** on
5 for	**6** with
7 about	**8** in

Reported speech

> **Study Guide** (pages 90–91)

Do you remember?

- This information reminds learners how to construct sentences that are in reported speech.
- Ask volunteers to read the advice given in speech bubbles, and then work through the information in the table. Make sure that learners understand the difference between statements and commands, and the subsequent differences in reported speech in each case.

Practise

- This task revises and strengthens learners' ability to construct reported speech in English. They read five speech bubbles and write sentences to report what each person said.
- You may wish to ask volunteers to read the speech bubbles aloud before learners begin this task.

Answers

1 She said that it was too late to go to the shops.

2 He said that they were planting a vegetable garden.

3 She said that Tim and Lara enjoy playing computer games.

4 May said that she likes to walk to school with her friends.

5 Dad shouted that I should come back.

Try this

- Put learners into pairs. Learners ask their partner to describe the weather and then write a sentence using reported speech to report what their partner said.
- This could be extended by repeating the exercise several more times with different partners.
- It is also a good opportunity to do some vocabulary revision relating to the weather, weather monitoring and recording equipment, climate zones, or seasons.

Answers

Learners' own responses, for example: My partner said that it is a beautiful day / is raining outside / is freezing cold and snowing!

Modals

Study Guide (pages 91–92)

Do you remember?

- Modal forms occur frequently in English sentences and they have a variety of uses, as described in this information.
- When learners are revising modal forms, pay close attention to those who tend to always use the same constructions; for example, they may always use 'can' and never 'will', or always use 'would' but never 'could'. Encourage these learners to remember that there are other modal forms, and to consciously try to use alternatives whenever possible.

Practise

This task reinforces the correct use of modal forms in sentences. Learners choose from three options to complete the sentences, with each option being a different modal form. They may find it helpful to say the sentences aloud and think about which modal 'sounds' and 'feels' right when making their their choices.

Answers

1 would

2 has to

3 should

4 ought not to

Try this

In this activity there is a dialogue with missing sentences. Learners need to decide which sentence belongs in which gap in the dialogue. Extend this activity by asking learners to perform the dialogue with a partner.

Answers

Yohan: Hi Marli. I need some advice please. Can you tell me what to do?

Marli: Sure. I'll try!

Yohan: I'm going to a new friend's house for supper. **Do I have to take a present?**

Marli: You needn't worry about a present. **But you should take a bunch of flowers.**

Yohan:: Flowers? OK. That's easy. Thanks.

Marli: But **you mustn't forget to send a message to say thank you.**

Adjectives and order of adjectives

Study Guide (pages 92–93)

Learn

- Ask the class to tell you what adjectives are and to give you some examples. Adjectives, put simply, are describing words. This information describes how some adjectives are formed by adding the '-ing' or '-ed' ending, as with present and past participles.
- Ask volunteers to read the examples in the bullet points and remind learners that important spelling rules apply when adjectives are formed in this way. Can learners recall what those spelling rules are? (Drop the final '-e' and double certain consonants.)

Practise

This task revises and practises forming adjectives by adding '-ing' or '-ed'. Remind learners to think about those important spelling rules!

Answers

1 excited, exciting   2 frightened, frightening

3 confused, confusing   4 bored, boring

Try this

- Learners have practised forming adjectives ending in '-ing' or '-ed', but do they understand the differences in meanings? As the *Hint* box states, these adjectives do not mean the same thing, so they are not interchangeable in sentences.
- There are three pairs of sentences for learners to consider in this activity. In each pair, learners must decide which is the correct sentence; in other words, is the sentence correct with an '-ing' adjective, or with an '-ed' adjective?

Answers

1 I was very bored during the movie.

2 My mum is interested in racing cars.

3 Are you frightened of dogs?

Learn

- It is possible to use more than one adjective in a sentence. In fact, we can use multiple adjectives in English sentences. However, when we do use two or more adjectives in a sentence, they must be placed in a particular order or the sentence won't make sense.
- Ask volunteers to read the speech bubbles, and then work through the order as explained in the table.
- Ask volunteers to make up some more examples following that pattern, for example 'three juicy red tomatoes'.

Try this

- This activity can result in a variety of different responses from learners and is an opportunity for them to be imaginative and creative. There are four items for learners to describe. Each item must be preceded by three adjectives. Learners can use the adjectives suggested in the Study Guide, or any other adjectives of their own choice.
- The key point here is that the adjectives are placed in the correct order in each case. There are many options for answers. If learners are in doubt about the validity of a response, they should check it in a dictionary.

Let's revise

- There will always be exceptions for every rule or pattern in grammar; we usually say that something is 'irregular' when this happens.
- Encourage learners to spend time focusing on irregular verbs, irregular adjectives or irregular spellings such as tricky plurals. Advise them to use mnemonics or 'look, cover, write, check' to help them to commit irregular forms to memory and to be able to recall them more quickly and easily.

Top tips

- When learned in isolation, grammar points can feel disjointed and remote from other parts of language learning. Try to pick up on grammatical concepts, to remind learners about grammar and to examine the grammar of English sentences at other times such as when reading texts or poems, when listening to audio recordings and during spoken activities.
- Grammar is not important only when writing; it is important to recognise and understand grammar when we read it, hear it and say it too!

Let's go! Worksheets 21–22: Grammar revision 1–2

- There are two grammar revision worksheets that learners can complete for additional support and reinforcement of grammatical concepts.
- Worksheet 21 revises the passive voice, past tense, present perfect tense, making suggestions and asking questions.
- Worksheet 22 revises first conditionals, talking about the future, prepositional phrases and reported speech.

Worksheet 21 answers

1 a At the Olympic Games, medals are won by athletes / Medals are won by athletes at the Olympic Games.

   b Fish are fed to the penguins by the zookeeper.

   c In an emergency, people are helped by firefighters / People are helped by firefighters in an emergency.

2 a watched

   b was playing

   c was writing

3 a have

   b have

   c has

   d have

   e has

   f has

4 Learners' own responses, for example:

Situation	Question or suggestion
Something to do on Sunday afternoon	We could go to the park.
You want to find out about buses to the city	What time is the bus to the city?
Somebody who could help solve a problem	How about Mum?
You want to know where something is located	Where is the cinema?
Something you would say in class	Can you help me, please?

Worksheet 22 answers

1 a It will get done quicker if we work together.

   b We can go to the beach if the weather stays dry.

   c If he scores this goal, the team can win the trophy.

2 a are watching

   b will play

   c will write

3 a for

   b with

   c about

   d of

   e at

4

Quote and speaker	Reported speech
'It was the worst storm for twenty years' the newsreader said.	The newsreader said that it was the worst storm for twenty years.
'The castle tour was very interesting' Dave said.	Dave said that the castle tour was very interesting.
'Put down the sweets, brush your teeth and go to bed!' Mum shouted.	Mum shouted at me to put down the sweets, brush my teeth and go to bed.
'How much does a karate lesson cost?' my brother asked	My brother asked how much a karate lesson costs.

Worksheet 21

Grammar revision 1

1 Change these sentences into the passive voice.

    a Athletes win medals at the Olympic Games.

    _____

    b The zookeeper feeds fish to the penguins.

    _____

    c Firefighters help people in an emergency.

    _____

2 Circle the correct word or words to complete each sentence

    a We (watched / was watching / watch) a movie last night.

    b The team (plays / playing / was playing) a match last Saturday.

    c I (write / was writing / writes) in my journal before I went to school.

3 Write **have** or **has** to complete each sentence.

    a We _____ been learning English for six years.

    b They _____ been to the theatre.

    c Paul _____ played guitar since he was ten.

    d Mum and Dad _____ asked us to tidy up.

    e The baby _____ been crying for hours!

    f Grandma _____ baked a surprise birthday cake for the party.

4 In the table, write a question or suggestion to match each situation.

Situation	Question or suggestion
Something to do on Sunday afternoon	
You want to find out about buses to the city	
Somebody who could help solve a problem	
You want to know where something is located	
Something you would say in class	

Worksheet 22

Grammar revision 2

1 Change these sentences to write the clauses the other way around.

    a If we work together, it will get done quicker.

    _____

    b If the weather stays dry, we can go to the beach.

    _____

    c The team can win the trophy if he scores this goal.

    _____

2 Circle the correct words to complete each sentence

    a We (watch / are watching / watches) a movie later.

    b The team (plays / will play / will playing) a match on Saturday.

    c I (writes / am writer / will write) in my journal after school.

3 Insert a preposition to complete each sentence.

    a What flavour juice did you ask _____?

    b Are you upset _____ me?

    c I care a lot _____ animals.

    d Dad has never heard _____ folktales!

    e I was laughing _____ the funny cartoon.

4 Convert each quote to reported speech. Write in the table.

Quote and speaker	Reported speech
'It was the worst storm for twenty years,' the newsreader said.	
'The castle tour was very interesting,' Dave said.	
'Put down the sweets, brush your teeth and go to bed!' Mum shouted.	
'How much does a karate lesson cost?' my brother asked	

Section 2: Vocabulary

 Study Guide pages 94–101

 Key skills: Using prepositions, adverbs and nouns correctly in sentences.

Prepositions

> **Study Guide** (pages 94–96)

Learn

- This section moves on from grammar to vocabulary. It would be a misconception for learners to think that vocabulary simply refers to lists of words grouped by topic areas and learned by rote. All words are part of our vocabulary. Section 2 explores words that have a variety of functions and begins here with prepositions that we use after adjectives.
- In Section 1, learners explored prepositions that we use to make prepositional phrases with verbs. Remind them of this and explain that we also make prepositional phrases with adjectives. The prepositions in the table are used in this way.
- Ask volunteers to give you examples of more prepositional phrases that they know, which could be added to those in the table, for example: 'scared of', 'happy with', 'known for', 'shocked at', 'worried about'. Be careful here that learners do not fall into the very easy trap of putting verbs before prepositions. A helpful hint for checking is to try putting the verb 'to be' before the adjective. If 'to be' works before the prepositional phrase, then it is an adjective; if you can't use 'to be' before the phrase, then it is probably already made of a verb. Praise learners for good effort, explain the mistake and ask them to try again.

Practise

This activity consolidates understanding of prepositional phrases with adjectives. Learners choose the correct preposition from three options to complete the sentences. Tell learners that they can refer back to the table in the *Learn* box on page 94 in the Study Guide if they are not sure.

Answers

1 of **2** at **3** of **4** about

Learn

- Another way that we can make prepositional phrases in English is by adding prepositions to nouns. This type of prepositional phrase usually describes a location or time, as shown by the examples in the table. We most commonly hear and use them when we are talking about travel and transport or events.
- Ask volunteers to read the examples in the table and then ask learners to give you more examples of their own to add to those in the table. After adding examples, refer to the *Hint* box, which advises learners that we can use concrete or abstract nouns with prepositions.
- Revise the difference between concrete and abstract nouns to make sure learners are secure before moving on.

Practise

This is a gap-fill exercise. Learners read six sentences and complete them with a preposition. Less-confident learners could work with a partner for peer support. Learners could swap with a partner and mark each other's work.

Answers

1 in **2** in **3** for **4** on **5** by, in **6** by

Do you remember?

- Another set of prepositions that learners know about are prepositions of manner. This information revises how the prepositions of manner 'like' and 'as' are used.
- Ask volunteers to read Sanchia's speech bubbles, then ask volunteers to give you some more examples of their own.

Try this

In this activity, learners fill the gaps in sentences to complete a dialogue. Learners could work with a partner to complete the activity and then they could perform the dialogue for the class or in small groups.

Answers

Mary: Are you going swimming?

Pam: No. You swim **like** a fish! I don't.

Mary: There are lots of people **like** you who can't swim.

Pam: Can you teach me how?

Mary: Yes. Soon you will love swimming **as** much **as** I do.

Adverbs

Study Guide (pages 96–98)

Learn

- Adverbs tell us more about verbs. There are different types of verbs, and this information looks at adverbs of time, which we use to describe when and how often things happen.
- Ask volunteers to read the example sentences in the bullet points.

Practise

There are eight adverbs in lozenges at the start of this activity. Learners need to rewrite sentences and include an adverb to give more detail. Learners can use any of the adverbs in any of the six sentences. Tell learners to think carefully about where to place the adverb in each sentence.

Answers

Learners' own responses, for example:

1 What are you doing **now**?

2 My grandmother **often** has a rest in the afternoons.

3 We are tested **monthly** to check how much we have learned.

4 I saw her **yesterday** when I went to the club.

5 Jo is excited because she is going to Paris **soon**.

6 They **usually** visit their cousins on weekends

Try this

- Learners read a recipe for a fresh salad. Then they rewrite the recipe and improve it by using adverbs of time to make the instructions clearer.
- The *Hint* box offers support and tells learners some useful adverbs they could put in the recipe instructions.
- You could extend this activity by having learners write their recipes on paper, adding a list of ingredients and including pictures. You could then make a colourful classroom display and perhaps link language revision to other topics such as healthy diet and lifestyle.

Do you remember?

When we want to make comparisons, we use comparative and superlative adverbs, as explained here. The table revises the rules for forming comparatives and superlatives and provides examples. Ask volunteers to give you other examples of their own, which you could list on the board.

Practise

This activity strengthens understanding of comparative and superlative adverbs. Learners choose the correct adverb from three options to complete sentences. Tell learners that they can refer back to the table in the *Do you remember?* box on page 97 in the Study Guide (or look at the list on the board) if they are not sure.

Answers

1 faster   2 beautifully   3 the worst   4 more slowly

Try this

- This is a picture-based comprehension activity. Learners look at the times on three digital watches belonging to a woman, a man and a boy, and then answer questions that relate to the three people.
- The advice in the *Hint* box reminds learners about when to use comparative and superlative forms in sentences.
- For the sentences in question 3, challenge more confident learners to compare the three people in their sentences, while less-confident learners could make comparisons between two people.

Answers

1 The man

2 Later

3 a The woman arrived <u>later than the boy</u>. (And the boy arrived later than the man.)

 b The boy arrived <u>earlier than the woman</u>. (And the man arrived earlier than the boy.)

Nouns

Study Guide (pages 98–101)

Do you remember?

- Learners know that nouns can be categorised in many different ways, and sometimes nouns can be in more than one group or category.
- Read, or ask volunteers to read, the information about different groups of nouns and ask learners to give examples of nouns that fit each group.

Learn

- One particular category is collective nouns. Emphasise that collective nouns are singular, as some learners find this difficult to understand since in a group there is more than one!
- Explain that no matter how many are in a group, there is still just one group; it is the group that is singular and that is why we use singular collective nouns.

Practise

How well can learners recall collective nouns? Without any additional resources, learners need to decide which of three collective noun options is correct in each sentence. Then they check their answers using a dictionary.

Answers

1 pack

2 fleet

3 swarm

4 band

Try this

Four different nouns are given in lozenges. The challenge is for learners to find the collective nouns for two of them – they may even already know some of these without having to use a dictionary. You could ask more confident learners to find the collective nouns for all four words.

Answers

A **herd** of elephants, a **team** of football players, a **school** of fish, a **choir** of singers.

Practise

- To engage learners with abstract nouns, this writing activity asks learners to describe themselves. They need to choose abstract nouns either from the box or others that they know, and they use them to describe their own character.
- Learners could share and compare their choices with a partner or in small groups. They could say whether they agree or disagree and offer alternative choices.

Try this

This crossword puzzle revises abstract nouns and their meanings. Nine abstract nouns are given in lozenges, these are the answers to the clues. Learners work out which abstract noun answers each clue and then they write the words in the correct place in the crossword grid.

Answers

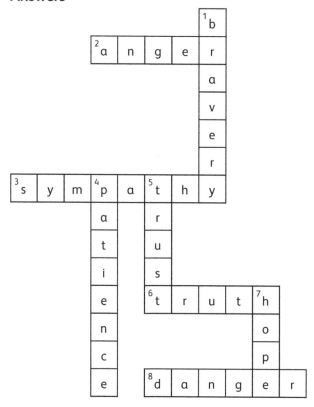

Learn

- Compound nouns can be tricky, and even home language English speakers find them problematic in writing. The reason for this is that we form compound nouns in several ways and, in some cases, more than one way of forming a compound noun is an acceptable spelling in English. The best thing to do is to learn compound nouns on a case-by-case basis.

- Ask volunteers to read the examples in the lozenges and to give more examples of compound nouns that they know.

Practise

This activity gives learners additional practice in using compound nouns. Each sentence has three compound noun options, but only one is spelled correctly. Learners should think carefully and then choose which one is correct.

Answers

1 Fireworks

2 great-grandmother

3 passers-by

4 earthquake

5 cupful

6 lifeguards

Try this

How well can learners interpret definitions? These four sentences have a word replaced with its definition (in bold for clarity). Learners rewrite the sentence, remove the definition and insert the correct word in its place. Each missing word happens also to be a compound noun.

Answers

1 surfboard

2 newspaper

3 facemasks

4 pop festival

Let's revise

- Revise numbers up to 1000 in a variety of ways, such as playing games of writing calculations in words and chanting or singing.
- Revise asking about time and telling the time using both digital and analogue clocks, and watches. You could do this by showing learners different times on clocks and watches, and asking them to write or say the times. Include revision of times of the day, parts of the day and periods of time such as a week, month, year and so on.

Top tips

Go back to the recipe writing activity on page 97 in the Study Guide and ask learners what other adverbs they could add in order to talk about safety. What sentences could they add to the recipe that relate to staying safe in the kitchen? Extend this into a discussion and encourage learners to give their opinions.

Let's go! Worksheets 23–24: Vocabulary revision 1–2

- Learners can complete two vocabulary revision worksheets for additional support and consolidation.
- Worksheet 23 revises prepositions with adjectives, as well as nouns, adverbs and adjectives.
- Worksheet 24 revises nouns and numbers.

Worksheet 23 answers

1 a now	**b** often	**c** never	**d** usually	**e** Sometimes	
2 a in	**b** on	**c** of	**d** by	**e** at	**f** after
3 a taller, tallest	**b** younger, youngest	**c** most beautiful	**d** curlier		
4 a faster	**b** earliest	**c** latest			

Worksheet 24 answers

1 Learners' own responses. There are many answer options. If you are in doubt about the validity of a response, check it in a dictionary. These are examples.

Category	Nouns
Countable nouns	apples, people, toys
Uncountable nouns	water, milk, sunshine
Concrete nouns	table, house, garden
Abstract nouns	love, fear, hope
Compound nouns	seahorse, earthquake, newspaper
Collective nouns for people	choir, team, band
Collective nouns for animals	pride, flock, swarm
Collective nouns for things	fleet, shower, pair

2 **a** 14   fourteen
  b 40   forty
  c 250   two hundred and fifty
  d 727   seven hundred and twenty-seven
  e 1000   one thousand
  f 90   ninety
  g 501   five hundred and one

Worksheet 23

Vocabulary revision 1

1 Circle the correct word to complete each sentence.

    a What are you doing (yesterday / sometimes / now)?

    b How (never / tomorrow / often) do you play tennis?

    c I (monthly / never / soon) arrive late for school.

    d We (usually / now / soon) go to the market at the weekend.

    e (Sometimes / Then / Annually) we have a takeaway for dinner.

2 Insert a preposition to complete each sentence.

    a I'm very interested _____ dinosaurs!

    b Uncle Pete is visiting _____ Friday.

    c I am very proud _____ myself.

    d We are going _____ plane.

    e She is quite good _____ baking cakes.

    f Let's go to the shops _____ school.

3 Insert the comparative and/or superlative form of the adjective in brackets to complete each sentence correctly.

    a My sister is _____ than me, but our Dad is the

    _____ in the family. (tall)

    b I am _____ than my brother, but my sister is the

    _____ of the children. (young)

    c I think the pink roses are the _____ flowers in the garden.
    (beautiful)

    d Your hair is _____ than mine. (curly)

4 Insert the comparative and/or superlative form of the adverb in brackets to complete each sentence correctly.

    a I can run _____ than all my friends (fast)

    b Michelle arrived first, so she was the _____. (early)

    c Matty arrived last, so he was the _____. (late)

Worksheet 24

Vocabulary revision 2

1 Look at the table. For each category, write at least three different nouns of your own choice. You may repeat a noun in more than one group if you need to, but try not to, if possible.

Category	Nouns
Countable nouns	
Uncountable nouns	
Concrete nouns	
Abstract nouns	
Compound nouns	
Collective nouns for people	
Collective nouns for animals	
Collective nouns for things	

2 Write each number in words.

a 14 _____

b 40 _____

c 250 _____

d 727 _____

e 1000 _____

f 90 _____

g 501 _____

Section 3: Sentence structure

 Study Guide pages 102–109

 Key skills: Constructing English sentences in a range of ways and by following patterns.

Quantifiers

Study Guide (pages 102–103)

Do you remember?
- This information revises how and why we use quantifiers in English sentences. It important for learners to be secure with recognising countable and uncountable nouns because they take different quantifiers.
- If you think it would be beneficial, spend some time reinforcing countable and uncountable nouns before moving on. Ask volunteers to read the example sentences in the bullet points.

Practise
- This is a multiple-choice activity where learners select the correct quantifier to complete five sentences.
- Ask learners to think about whether the noun is countable or uncountable, or if a comparison is being made.

Answers

1 A   2 C   3 A   4 B   5 C

Try this
- This activity involves imagination and creative thinking as learners imagine they are shopping. They complete and continue a short dialogue between the shop assistant and a customer. The conversation naturally includes many quantifier words as the customer compares items and asks about prices.
- Learners can then perform their dialogues for the class or in small groups. Answers for the gaps in the first half of the dialogue are provided below. The rest of the dialogue will vary.

Answers

Customer: Hello. Do you have **any** sun hats?

Assistant: Yes, we have **plenty / some** to choose from.

Customer: Oh good. Those look a bit small, but I like these two styles. Do you have **either** in a bigger size?

Assistant: Yes, we have **both**.

Customer: Great. How **much** are they?

Pronouns

Study Guide (pages 103–105)

Learn
- There are several different types of pronouns. Learners will be most familiar with the personal pronouns that we use with verbs and in place of nouns in sentences. Other types of pronouns include demonstrative pronouns and indefinite pronouns, which are explained here.
- Ask volunteers to read the example sentences. Use the labelled pictures to show learners exactly what the pronouns in each sentence refer to.

Practise
Learners read six sentences and identify the pronouns. Ask learners to read slowly and carefully because there can be more than one pronoun in a sentence.

Answers

1 those	2 He, I	3 you, me
4 Everyone	5 We, something, nothing	6 That, I, it

Try this (page 104)

- Learners can work on this activity with a partner or individually. They need to complete a dialogue by choosing from the sentences provided to fill in the missing lines.
- With a partner, learners could read the dialogue aloud twice, swapping roles so that they each read both parts.

Answers

Andile: Which jeans should I buy?

Thabo: How about these?

Andile: But these have holes in them!

Thabo: Yes, but holes are trendy!

Andile: They look old.

Thabo: These are cool. Do you like them?

Andile: No. I don't like black jeans.

Thabo: Don't you like anything in this shop?

Andile: No, not really. It's too expensive.

Learn

Here, learners are reminded about reflexive pronouns, which always end in '-self' or '-selves'. Read each pronoun from the lozenges and have learners repeat it back to you. It can be helpful to point or use arm gestures for reinforcement. For example, point at yourself when you say 'myself', make a circling gesture as if including everyone when you say 'ourselves'.

Practise

Learners complete gaps in sentences by inserting a reflexive pronoun. Advise learners to identify who is being referred to, and to look for other pronouns in the sentence to help them decide which reflexive pronoun they need to use.

Answers

1 myself	2 yourself	3 herself
4 ourselves	5 themselves	6 himself

Try this

- Here, learners read sentences that each have a deliberate mistake. The pronouns in bold text are incorrect. Learners need to rewrite the sentences and include the correct pronouns.
- They could swap with a partner, mark each other's work and provide feedback.

Answers

1 They took me home with **them**.

2 They gave **him** a prize for working hard.

3 I have two bags. Both of **these** are mine.

4 Jenna and I went to a concert.

Clauses and connectives

> **Study Guide** (pages 105–108)

Do you remember?

- This information revises how clauses are used to structure sentences, and how the clauses are most often joined together by connectives.

- Ask a volunteer to read Maris's speech bubble about the difference between clauses and phrases. Then ask learners to read the information about connectives and allow time for them to ask questions.
- There are three sentences in bullet points at the end. Tell learners to identify the two clauses in each sentence. It may be helpful to write the sentences on the board and use different colour pens for each clause or separate the clauses with a slash for emphasis.

Practise

- This activity reinforces connectives. Learners choose the most appropriate connective to join the clauses in the four sentences.
- Refer learner to the *Hint* box, which advises them to make sure they understand the sentences before they choose the connectives.

Answers

1 as soon as   **2** until   **3** then   **4** as

Try this

Learners should try to complete the first part of this activity individually, and then in the second part of the activity, they team up with a partner. First, they rewrite several short sentences by joining some of them together using connectives. Then they compare their choices with a partner and discuss them.

Answers

1 We waited at the bus stop **until** the bus came.

We ate our lunch on the bus, **then** we slept.

We arrived at my aunt's house six hours later.

Do you remember?

- Some clauses in English sentences can be joined together using other words, specifically the 'wh' words. This information revises how we use these words to give information about people or things. Ask volunteers to read the examples in the speech bubbles.
- The information then goes on to describe how our choice of which word to use can depend on context, because certain words are more formal. This is quite a subtle distinction that depends somewhat on context, so learners may need support with recognising situations in which the more formal tone would be appropriate.

Practise

- The four sentences in this activity reinforce the use of 'wh-' words before clauses. Learners choose which word is the correct option to complete the sentence in each case.
- For extra support, go through the answers with the class, examining each sentence and clarifying the reasons for choosing each word.

Answers

1 who   **2** that   **3** to whom   **4** whose

Try this

This writing activity has two stages. To provide differentiated support for less-confident learners, put learners into pairs or small groups first to talk through this activity together. They can discuss which parts of the sentences could match, perhaps by saying them aloud with different parts or writing them on rough paper. When they are happy, learners can then write out the final sentences neatly in the space provided.

Answers

1 Paulo is part of our team whom you know already.

2 The music was very loud, which made me feel deaf.

3 Is this the boy whose shoes you borrowed?

4 A pilot is a person who flies planes and helicopters.

5 Let's go to the park that I was telling you about.

Learn

- Some types of clauses are called subordinate clauses. This information revises how these clauses are used in reported speech and that we introduce subordinate clauses using the verbs 'say' or 'tell'.
- Ask volunteers to read the example sentences in the bullet points, and then read Sanchia's speech bubble yourself. Ask her question directly to your class and encourage them to look at the pronouns and the word order. Remind learners to use 'that' after these verbs; otherwise the sentences will not make sense.

Practise

- This activity will reinforce and consolidate how to use subordinate clauses in reported speech. Learners rewrite four quotes as reported speech. They do this twice for each quote, to practise using both verbs 'say' and 'tell'.
- Remind learners to include 'that' when they rewrite their sentences and to think carefully about word order.

Answers

1 The baker said that / told us that we should mix the flour and the butter first.

2 The shop assistant said that / told us that the shop closes at 5:30.

3 He said / He told us that we should try to eat fruit and vegetables every day.

4 Petra said to Milly that / Petra told Milly that Jo was quite angry because she didn't invite her.

Other sentence patterns

> **Study Guide** (pages 108–109)

Do you remember?

There are two other common sentence structures in English that learners should recognise. The first of these, discussed here, is the 'verb + object + infinitive' structure. There are some exceptions for cases in which we can't use this structure. Ask volunteers to read the example sentences in the bullet points, which explain this.

Practise (page 108)

This revision exercise reinforces sentences with infinitives. Learners choose the correct word to complete four sentences that follow the pattern.

Answers

1 me   2 to sit   3 move   4 to help

Do you remember?

The second sentence structure, discussed here, is when we make sentences using verbs and objects. Again, ask volunteers to read the example sentences in the bullet points and then move on to the revision exercises.

Practise

This activity will test learners' ability to recognise sentences that are grammatically correct. There are seven pairs of sentences, but only one sentence in each pair is written in correct English. How many correct sentences can learners identify?

Answers

1 My mother showed us the jumper that she knitted.

2 'Can you bring me some water, please?' asked my aunt.

3 Marja sends her grandmother a gift for her birthday every year.

4 Do I have to do this work today or can I do it tomorrow?

5 Will you take this parcel to the Post Office please?

6 She told me to give her some advice.

7 The boy kicked a ball as he walked along the road.

Try this

- This writing activity could provide a variety of responses as learners construct their own sentences using five given sentence stems. When learners have written their sentences, put them into small groups to share and compare their work. How similar and different are their sentences?
- There are many options for answers. If learners are in doubt about the validity of a response, they should check it in a dictionary.

Let's revise

Use reading texts for revision and reinforcement of clauses. Learners could take a page or a paragraph from a story or other reading material and use a highlighter to define the clauses in the sentences.

Top tips

You can revise sentence structures by giving learners sentences where the words are jumbled and asking them to rewrite the sentences using every word. Learners will need to look for key words such as participles and verb forms to help them identify the type of sentence in each case.

Let's go! Worksheets 25–26: Sentence structure revision 1–2

- There are two sentence structure revision worksheets that learners can complete for additional support, reinforcement and consolidation.
- Worksheet 25 revises quantifiers, pronouns and connectives.
- Worksheet 26 revises clauses and other sentence patterns, before giving learners opportunity to complete a self-reflection exercise, which enables them to think about what they are most confident with, what they need to work on some more and what they need extra help with.

Worksheet 25 answers

1 a pinch	**b** both	**c** much	**d** many	**e** neither	
2 a She	**b** this / that	**c** those / my	**d** enough	**e** each	**f** everyone
3 a herself	**b** myself	**c** yourself / yourselves	**d** ourselves		
4 a before	**b** after	**c** until	**d** because	**e** and	**f** but

Worksheet 26 answers

1 a who	**b** which	**c** where	**d** that	**e** whose
2 a to put	**b** to buy	**c** to read	**d** to catch	**e** to win
3 a showed	**b** bring	**c** taking / take	**d** send	**e** brought

Worksheet 25

Sentence structure revision 1

1 Circle the correct quantifier to describe each noun.

   a A (pinch / litre) of salt

   b (Both / either) teams play in a red kit

   c I don't want too (much / many) ketchup, thank you.

   d There are so (much / many) options to choose from.

   e There was no winner because (neither / either) team scored any points.

2 Fill the gaps to complete each sentence.

   a _____ likes to visit her gran every weekend.

   b Is _____ the one you wanted?

   c Where did I put _____ keys?

   d Do you drink _____ water every day?

   e Share the sweets equally, so you get five _____.

   f Quiet, _____, please!

3 Insert a reflexive pronoun to make each sentence correct.

   a My baby sister can crawl by _____.

   b I want to try to finish the project _____.

   c You can watch the movie _____ after school.

   d We pride _____ on always being polite and helpful.

4 Insert a connective to make each sentence correct.

   a Heat the frying pan _____ you pour in the mixture.

   b I found out about the earthquake _____ it had happened.

   c Can you stay _____ the end of the party?

   d They need to wait there _____ the door is locked.

   e Choir practice is on Tuesday _____ Thursday this week.

   f Yes, you can go to the park, _____ wear a sun hat!

Worksheet 26

Sentence structure revision 2

1 Choose the best word to connect the clauses in each sentence.

   a   An artist is a person (who / which) draws or paints.

   b   A helmet is an item (which / whom) protects the head.

   c   A hospital is a place (that / where) patients are cared for.

   d   The new shopping mall is the one (where / that) we are visiting later.

   e   The team (whose / who) score is the lowest will lose the match.

2 Underline the infinitive in each sentence.

   a   I told you to put it away.

   b   Grandpa asked me to buy him a newspaper.

   c   My teacher wants me to read the poem aloud.

   d   We waited at the corner to catch the bus.

   e   He ran quickly to win the race.

3 Insert the correct part of the verb **give**, **take**, **send**, **bring** or **show** to make each sentence correct.

   a   Pam _____ us the cakes she had baked.

   b   If you _____ the bag to me, I will carry it home.

   c   A witness saw the thief _____ the jewels.

   d   The best way to stay in touch is to _____ an email.

   e   Lucy _____ her pet tortoise to school today for show and tell.

Self-reflection

What am I confident about?
What do I need more practice with?
What do I need more help with?

Check your understanding

Study Guide pages 110–111

- Pages 110 and 111 in the Study Guide have some end-of-unit exercises that can be used to check how well learners understand the grammar that has been covered throughout Unit 4.

- There are four tasks in total. The first is a multiple-choice exercise where learners complete sentences by choosing the correct word from a choice of three. Then learners complete a dialogue, which is followed by completing gaps in an email. There are no options provided in the email exercise, so learners need to read the whole sentence, think about context as well as sentence structure, and then decide which word to use for each gap.

- The final exercise is about self-reflection. Learners think about their progress throughout Unit 4 and they consider what they enjoyed most and what they need to spend more time on for revision.

Answers

1 **a** bitten
   b won't
   c walking
   d swarm
   e much
   f for
   g of
   h ourselves
   i them
   j myself
   k anything

2 **Anwar:** Do you know how to play cricket?
   Tim: Yes, I do! It's fun.
   Anwar: **Can** you show me how to play?
   Tim: Yes, of course.
   Anwar: **What** do we need to play?
   Tim: I have some cricket bats and balls we can use.
   Anwar: **Where** can we play?
   Tim: There is an open field near my flat. We can go there.
   Anwar: **Is** it difficult?
   Tim: No, it isn't. You have to learn some rules though.
   Anwar: **Who** taught you how to play?
   Tim: My dad and mum taught me! **Would** you like to try now?
   Anwar: Yes, please! Let's go.

3 **a** would   **b** for
   c when   **d** many
   e who   **f** what about
   g with   **h** you

4 Learners' own responses.

Hints for revision

Some useful hints and tips are presented on page 112 of the Study Guide to help learners as they are revising for the Cambridge Checkpoint tests. These tips have been written by the author.

- On the page there are four boxes, one for each of the key skills. You could photocopy the page and enlarge it to display the revision tips in the classroom.

- Spend some time in class discussing the advice. Ask learners to identify if they already follow any of the hints by asking questions such as: *Who here reads something in English every day?*

- Dig deeper by focusing on particular points from the revision tips. For example, one of the tips about writing is to watch out for homophones. Ask volunteers to give you examples of homophones. You could list them on the board, or you could put learners into small groups (teams) and see how many homophones each team can think of. Similarly, in the *Use of English* advice, there is a hint about using connectors. How many different connective words can the class recall?

- If you prefer, learners could discuss the revision tips in groups. Put learners into four groups and allocate one of the key skills to each group. Tell learners to read the advice and talk about it together. Do they already do these things? Is the advice helpful? Do they have any other advice relating to this skill that they could add to the tips? When learners have had time for discussion, ask each group to feedback to the class about the advice they have read and to share their ideas from the discussion.

- As a homework activity, you could ask learners to make their own *Hints for revision* poster. They can choose their favourite advice from the Study Guide and put it into a bright, colourful poster with their own added drawings to match. They might prefer to make a poster of their own advice for revision. They could then bring their posters to class and share and compare them with a partner or in small groups.

- The final piece of advice on page 112 of the Study Guide says 'Good luck! And remember – ALWAYS read instructions and questions carefully.' Be sure to emphasise this advice with the class. Reading instructions slowly and carefully is essential. Even if something looks straight forward, rushing in is never a good idea! Explain that mistakes are much more likely to happen if instructions are not read properly. This simple piece of advice could stop learners from missing out on lots of points in any future tests.

- You could also recommend that parents or carers read these tips and support learners to use some of the techniques that are described. For example, they could put aside some time every day for learners to read, or to listen to them reading aloud. They could encourage learners to practise speaking with other members of the family who can speak English.

Revision Practice Test instructions

A photocopiable *Revision Practice Test* is provided in the following section on pages 107–111. These questions have been written by the author of the Teacher's Guide. Answers to all questions in the test are provided later on page 112.

- There are 8 sections in the test, with 40 questions altogether. Questions are worth 1 mark each, so the test is marked out of 40. Each section has one main instruction followed by several numbered questions. Remind learners to always read the instructions in every section before they begin to answer the questions. Each section has its own point of focus as follows:

Section 1	Vocabulary and spellings
Section 2	Quantifier words
Section 3	Collective nouns
Section 4	Indirect speech
Section 5	Adjectives with **-ing** and **-ed**
Section 6	Present perfect tense
Section 7	Connective words and phrases
Section 8	Modal verbs

- There are a variety of question types including multiple-choice questions, gap-fill exercises and language recognition activities. A cover page is provided, which should also be photocopied and provided with the test. Learners write their name on the cover page and read the instructions before they complete the test.

- An approximate amount of time to allow for the test would be around 30 minutes, but you should extend or shorten this time according to the needs of the class.

- To provide differentiated support for less-confident learners, you may wish to choose which sections of the test they complete, or you could give them each section of the test separately over a few sessions.

- The *Revision Practice Test* can serve as a formative assessment. You can use the results to help you to see which learners need additional support and what they need support with. In this way, you can provide individual intervention and advice.

Revision Practice Test

Name: _____

Instructions

- There are 40 questions in total.
- Each question is worth 1 mark.
- Try your best to attempt all of the questions.
- Read the instructions for each question carefully.
- Write your answers on the paper.
- Use your best handwriting.

Cambridge Primary Revise for Primary Checkpoint World English Teacher's Handbook © Jennifer Peek 2022

Name: _____

Section 1
Questions 1–10

Find the correct spellings. For each question, choose the best answer: **A**, **B** or **C**. Circle your choice.

1	**A** sibllings	**B** siblings	**C** sibblings		

2   **A** shud   **B** shoud   **C** should

3   **A** profesor   **B** proffesor   **C** professor

4   **A** huricane   **B** hurricane   **C** hurracane

5   **A** thermometer   **B** thermometre   **C** thermomiter

6   **A** drouhts   **B** droughts   **C** drouhgts

7   **A** bllizzard   **B** blizard   **C** blizzard

8   **A** pilet   **B** pilot   **C** pilat

9   **A** meckanic   **B** meckhanic   **C** mechanic

10 **A** nurs   **B** nerse   **C** nurse

Section 2
Questions 11–14

Read the sentences. Circle the **quantifier** word(s) in each question.

11 Several people are waiting for the train.

12 There's plenty of fruit in the kitchen.

13 I want a job with very little stress.

14 Not many people like working at weekends.

Name: _____

Section 3
Questions 15–20

Tick (✔) the correct collective noun for each noun in the table.

		pack	pile	bunch
15	sand			
16	cards			
17	rubbish			
18	flowers			
19	wolves			
20	coconuts			

Section 4
Questions 21–23

Read the sentences. Rewrite sentences 21, 22 and 23 using indirect speech.

21 'Snow is expected to fall this weekend,' said the news reporter.

_____

_____

22 'Move people and animals upstairs to shelter from the flood,' she said.

_____

_____

23 'The storm is the worst in 20 years,' said the firefighter.

_____

_____

Name: _____

Section 5
Questions 24–28

Choose the correct word to complete each sentence. For each question, choose the best answer: **A**, **B** or **C**. Circle your choice.

24 Pia is very _____ in learning more about space.

    A interesting   **B** interest   **C** interested

25 David doesn't like snakes; he thinks they are _____.

    A frightened   **B** frightening   **C** frighten

26 Sanchia is _____ about the class trip to the aquarium.

    A exciting   **B** excited   **C** excite

27 Maris finds crossword puzzles quite _____.

    A confuse   **B** confused   **C** confusing

28 Banko _____ learning about the weather.

    A enjoy   **B** enjoyed   **C** enjoying

Section 6
Questions 29–32

Read the sentences. Complete each sentence using **for**, **since**, **never** or **ever**. Use each word only once.

29 Have you _____ hiked in a forest?

30 We are going camping _____ the first time.

31 I've _____ slept in a tent, but I'd like to.

32 We've been waiting to start the hike _____ 9 o'clock!

Name: _____

Section 7
Questions 33–36

Read the sentences. Choose the best **connective** to fill the gap. For each question, choose the best answer: **A**, **B** or **C**. Circle your choice.

33 _____, we need to look after the planet.

    A Such as   **B** Clearly   **C** Who

34 Solar power is good, _____ it isn't always sunny!

    A for example   **B** of course   **C** however

35 Wind turbines are a positive change but, _____, they are ugly.

    A on the other hand   **B** for example   **C** such as

36 We can save energy at home, _____ by turning off lights.

    A when   **B** for example   **C** clearly

Section 8
Questions 37–40

Complete each sentence by writing a modal verb. Choose **should**, **ought to**, **have to** or **mustn't**. Use each word only once.

37 We _____ use more renewable energy.

38 You can go to the park, but you _____ stay out after dark

39 That's a great idea! You _____ share it with the class.

40 I want to buy a new skateboard so I _____ save my pocket money.

Cambridge Primary Revise for Primary Checkpoint World English Teacher's Handbook © Jennifer Peek 2022

Revision Practice Test Answers

Section 1, Questions 1–10
1 B   2 C   3 C   4 B   5 A   6 B   7 C   8 B   9 C   10 C

Section 2, Questions 11–14
11 Several   12 plenty   13 (very) little   14 (Not) many

Section 3, Questions 15–20

		pack	pile	bunch
15	sand		✔	
16	cards	✔		
17	rubbish		✔	
18	flowers			✔
19	wolves	✔		
20	coconuts			✔

Section 4, Questions 21–23
21 The news reporter said that snow was expected to fall this weekend.
22 She said that we should move people and animals upstairs to shelter from the flood.
23 The firefighter said that it was the worst storm in 20 years.

Section 5, Questions 24–28
24 C   25 B   26 B   27 C   28 B

Section 6, Questions 29–32
29 Have you ever hiked in a forest?
30 We are going camping for the first time.
31 I've never slept in a tent, but I'd like to.
32 We've been waiting to start the hike since 9 o'clock!

Section 7, Questions 33–36
33 B   34 C   35 A   36 B

Section 8, Questions 37– 40
37 ought to / should
38 mustn't
39 should / ought to
40 have to